Operations Research/Computer Science Interfaces Series

Volume 53

For further volumes:
http://www.springer.com/series/6375

Luca Di Gaspero • Andrea Schaerf
Thomas Stützle
Editors

Advances in Metaheuristics

Springer

Editors
Luca Di Gaspero
DIEGM
University of Udine
Udine, Italy

Andrea Schaerf
DIEGM
University of Udine
Udine, Italy

Thomas Stützle
IRIDIA
Université Libre de Bruxelles
Bruxelles, Belgium

ISSN 1387-666X
ISBN 978-1-4614-6321-4 ISBN 978-1-4614-6322-1 (eBook)
DOI 10.1007/978-1-4614-6322-1
Springer New York Heidelberg Dordrecht London

Library of Congress Control Number: 2013931180

Printed on acid-free paper

Springer is part of Springer Science+Business Media (www.springer.com)

Preface

Metaheuristics have been a very active research topic for more than two decades. During this time many new metaheuristic strategies have been devised, they have been experimentally tested and improved on challenging benchmark problems, and they have proven to be important tools for tackling optimization tasks in a large number of practical applications. In other words, metaheuristics are nowadays established as one of the main search paradigms for tackling computationally hard problems. Still, there are a large number of research challenges in the area of metaheuristics. These challenges range from more fundamental questions on theoretical properties and performance guarantees, empirical algorithm analysis, effective configuration of metaheuristic algorithms, approaches to combine metaheuristics with other algorithmic techniques, towards extending the available techniques to tackle ever more challenging problems.

This edited volume grew out of the contributions presented at the ninth Metaheuristics International Conference that was held in Udine, Italy, 25–28 July 2011. The conference comprised 117 presentations of peer-reviewed contributions and 3 invited talks, and it was attended by 169 delegates. The chapters collected in this book exemplify contributions to several of the research directions that we have outlined above.

The first three contributions deal with the analysis and the configuration of metaheuristics. The article of Lehre and Witt presents first steps towards the theoretical run-time analysis of particle swarm optimization (PSO) algorithms, a well-known metaheuristic technique that is mainly used to tackle continuous optimization problems. In particular, they analyze PSO algorithms from the perspective of the first hitting time of a (small) region in the solution space around the optimum.

Barbosa, Bernardino, and Barreto deal with techniques for analyzing the run-time behavior of metaheuristics. They discuss variants of performance profiles, a tool for the visualization and the interpretation of experimental results, and they view performance comparisons from a multi-criteria perspective.

Gunawan, Lau, and Wong explore the possibility of using experimental design techniques to reduce the size of the parameter space that automatic algorithm configuration techniques need to consider. Essentially, they propose to avoid the combinatorial explosion of full factorial designs by working on partitions of parameters and then restricting the domains of the parameters using fractional factorial designs. In their contribution they illustrate their approach using one case study.

The following three chapters describe applications of metaheuristics to challenging problems. The inherent difficulty in these problems is due to the inclusion of multiple objective functions, dynamic data, or a mix of different types of variables. Wagner, Day, Jordan, Kroeger, and Neumann search for pacing strategies that improve the performance of a cycling team in team pursuit track cycling. The hardness of the problem arises from the simultaneous optimization of the sequence of actions the cyclists should do (i.e., permutations of integer values) and the power they should invest (i.e., continuous variables).

Amorim, Henggeler Antunes, and Almada–Lobo explore the integration of dual information into the mutation operator of a genetic algorithm. Their approach is applied to a challenging multi-objective optimization problem that arises in the production planning of perishable products. The main contribution of the paper is to show how to employ dual information from a mixed-integer programming formulation of the problem to better direct the mutation operator of a multi-objective genetic algorithm.

Lepagnot, Nakib, Oulhadj, and Siarry propose a new metaheuristic technique, called *multiple local search algorithm for dynamic optimization (MLSDO)*, which has been applied for the solution of an image registration problem. In detail, the system is designed to recognize abnormalities in the brain, starting from a sequence of cine-MRI registrations. MLSDO uses several local searches to explore the search space and to track the found optima over the dynamic modifications of the objective function. In this approach, each local search is performed in parallel, and all searches are coordinated by a dedicated module.

The final four chapters contribute new metaheuristic approaches to tackle difficult combinatorial optimization problems. The first two deal with vehicle routing problems, a relevant class of optimization problems where metaheuristics have reached particularly high success.

Crainic, Mancini, Perboli, and Tadei propose an approach based on the GRASP metaheuristic. In this work, GRASP is used in combination with path relinking to address the two-echelon vehicle routing problem. This problem is an extension of the classical vehicle routing in which the delivery from a single depot to customers is achieved by routing and consolidating the freight through intermediate depots, called satellites. The problem is treated by decomposition, separating the depot-to-satellite transfer from the satellite-to-customer delivery. The approach applies a GRASP and a local search procedure in sequence. Afterwards, the resulting solution is linked to an elite solution by means of path relinking.

Reinholz and Schneider tackle the open vehicle routing problem, whose main characteristic is that vehicles do not return to the depot after serving the last client. The authors define a framework that includes several standard neighborhood structures that are based on a path exchange operator and that allows for an efficient computation of cost differences induced by the moves. The algorithmic approach uses these neighborhood structures in a stochastic multiple neighborhood search and integrates the latter into a (1+1)-evolutionary strategy. The performance of this hybrid strategy is evaluated on standard benchmark instances taken from literature, proving the competitiveness of the method.

Pimmer and Raidl deal with another classical problem that proved to be particularly hard in practice, namely the high-school timetabling problem. The approach used in this chapter is not a standard metaheuristic approach, but a complex heuristic composed by a constructive phase followed by a local search. Both phases are applied at the level of groups of events in the same timeslot, in a large-neighborhood fashion. Experimental results on available benchmarks indicate that this approach is competitive.

Kemmoe, Lacomme, Tchernev, and Quilliot use GRASP to address a supply chain optimization problem with additional financial constraints. In this setting, the goal is to obtain the smallest duration of a supply chain operational planning that respects the budget limits. For this specific problem, GRASP is hybridized with an evolutionary local search. The method is evaluated on a set of widely used benchmarks, suitably modified to take into account the financial constraints.

The organization of MIC 2011 and the edition of this post-conference volume would not have been possible without the help of a large number of people. Special cordial thanks go to the members of the local organizing committee for their hard work in all preparatory tasks and the practical organization of the conference. We also would sincerely thank the special session organizers for extending the conference program with relevant topics that made the conference more attractive, as well as the program committee members and the additional referees for their qualified and detailed reviews. Last but not least, we would like to thank all the researchers in metaheuristics who submitted and participated in the conference and, in particular, who submitted their extended versions for consideration in this post-conference volume: you are the most important to advance the metaheuristics field and to make metaheuristic research lively, inspiring, and enjoyable!

Udine, Italy — Luca Di Gaspero
Udine, Italy — Andrea Schaerf
Bruxelles, Belgium — Thomas Stützle

Acknowledgements

Printed with assistance of the Department of Electrical, Management, and Mechanical Engineering of the University of Udine.

Contents

Finite First Hitting Time Versus Stochastic Convergence in Particle Swarm Optimisation

Per Kristian Lehre and Carsten Witt

Abstract We reconsider stochastic convergence analyses of particle swarm optimisation, and point out that previously obtained parameter conditions are not always sufficient to guarantee mean square convergence to a local optimum. We show that stagnation can in fact occur for non-trivial configurations in non-optimal parts of the search space, even for simple functions like SPHERE. The convergence properties of the basic PSO may in these situations be detrimental to the goal of optimisation, to discover a sufficiently good solution within reasonable time. To characterise optimisation ability of algorithms, we suggest the expected first hitting time (FHT), i.e., the time until a search point in the vicinity of the optimum is visited. It is shown that a basic PSO may have infinite expected FHT, while an algorithm introduced here, the Noisy PSO, has finite expected FHT on some functions.

1 Introduction

Particle Swarm Optimisation (PSO) is an optimisation technique for functions over continuous spaces introduced by Kennedy and Eberhart [9]. The algorithm simulates the motions of a swarm of particles in the solution space. While limited by inertia, each particle is subject to two attracting forces, towards the best position P_t visited by the particle, and towards the best position G_t visited by any particle in the swarm. The update equations for the velocity V_t and the position X_t are given in Algorithm 1 in Sect. 2. The inertia factor ω, and the acceleration coefficients φ_1 and φ_2 are user-specified parameters. The algorithm only uses objective function

Carsten Witt
DTU Informatics, Technical University of Denmark, Lyngby, Denmark
e-mail: cfw@imm.dtu.dk

Per Kristian Lehre
School of Computer Science, University of Nottingham, Nottingham, UK
e-mail: PerKristian.Lehre@nottingham.ac.uk

L. Di Gaspero et al. (eds.), *Advances in Metaheuristics*, Operations Research/ Computer Science Interfaces Series 53, DOI 10.1007/978-1-4614-6322-1_1,

values when updating G and P, and does not require any gradient information. So the PSO is a black-box algorithm [4]. It is straightforward to implement and has been applied successfully in many optimisation domains. Despite its popularity, the theoretical understanding of the PSO remains limited. In particular, how do the parameter settings influence the swarm dynamics, and in the end, the performance of the PSO?

One of the best understood aspects of the PSO dynamics are the conditions under which the swarm stagnates into an equilibrium point. It is not too difficult to see (e.g., [3]) that velocity explosion can only be avoided when the inertia factor is bounded by

$$|\omega| < 1. \tag{1}$$

The magnitude of the velocities still depends heavily on how the global G_t and local best P_t positions evolve with time t, which again is influenced by the function that is optimised. To simplify the matters, it has generally been assumed that the swarm has entered a stagnation mode, where the global and local best particle positions G_t and P_t remain fixed. Under this assumption, there is no interaction between the particles, or between the problem dimensions, and the function to be optimised is irrelevant. The swarm can therefore be understood as a set of independent, one-dimensional processes.

An additional simplifying assumption made in early convergence analyses was to disregard the stochastic factors R and S, replacing them by constants [13, 3]. Trelea [12] analysed the one-dimensional dynamics under this assumption, showing that convergence to the equilibrium point

$$P_e = (\varphi_1 P + \varphi_2 G)/(\varphi_1 + \varphi_2) \tag{2}$$

occurs under condition (1) and

$$0 < \varphi_1 + \varphi_2 < 4(1+\omega). \tag{3}$$

Kadirkamanathan et al. [8] were among the first to take the stochastic effects into account, approaching the dynamics of the global best particle position (for which $P = G$) from a control-theoretic angle. In particular, they considered asymptotic Lyapunov stability of the global best position, still under the assumption of fixed P and G. Informally, this stability condition is satisfied if the global best particle always converges to the global best position when started nearby it. Assuming a global best position in the origin, their analysis shows that condition (1), $\omega \neq 0$, and

$$\varphi_1 + \varphi_2 < 2(1 - 2|\omega| + \omega^2)/(1+\omega) \tag{4}$$

are sufficient to guarantee asymptotic Lyapunov stability of the origin. These conditions are not necessary, and are conservative. Another stochastic mode of convergence considered, is convergence in mean square (also called second order stability) to a point x^*, defined as $\lim_{t\to\infty} E\left(|X_t - x^*|^2\right) = 0$. Mean square convergence to x^*

implies that the expectation of the particle position converges to x^*, while its variance converges to 0. It has been claimed that all particles in the PSO converges in mean square to the global best position if the parameter triplet $\omega, \varphi_1, \varphi_2$ is set appropriately. Jiang et al. [7] derived recurrence equations for the sequences $E(X_t)$ and $\mathrm{Var}(X_t)$ assuming fixed G and P, and determined conditions, i.e. a convergence region, under which these sequences are convergent. The convergence region considered in [7] is strictly contained in the convergence region given by the deterministic condition (3). For positive ω, the Lyapunov stability region described by condition (4) is strictly contained in the mean square stability region. Given the conditions indicated in Fig. 1, the expectation will converge to P_e [as in Eq. (2)], while the variance will converge to a value which is proportional to $(G-P)^2$. It is claimed that the local best P converges to G, which would imply that the variance converges to 0. However, as we will explain in later sections, this is not generally correct. We will discuss further assumptions that are needed to fix the claim of [7]. Wakasa et al. [14] pointed out an alternative technique for determining mean square stability of specific parameter triplets. They showed that this problem, and other problems related to the PSO dynamics, can be reduced to checking the existence of a matrix satisfying an associated linear matrix inequality (LMI). This is a standard approach in control theory, and is popular because the reduced LMI problem can be solved efficiently using convex- and quasi-convex optimisation techniques [1]. Wakasa et al. [14] obtained explicit expressions for the mean square stability region, identical to the stability region obtained in [7], using this technique. Assuming stagnation, Poli [10] provided recurrence equations for higher moments (e.g. skewness and kurtosis) of the particle distribution. The equations for the m-th moment are expressed with an exponential number of terms in m, but can be solved using computer algebra systems for not too high moments.

Recently, there has been progress in removing the stagnation assumption on P and G. Building on previous work by Brandstätter and Baumgartner [2], Fernández-Martínez and García-Gonzalo [5] interpret the PSO dynamics as a discrete-time approximation of a certain spring-mass system. From this mechanical interpretation follows naturally a generalisation of the PSO with adjustable time step Δt, where the special case $\Delta t = 1$ corresponds to the standard PSO. In the limit where $\Delta t \to 0$, one obtains a continuous-time PSO governed by stochastic differential equations. They show that dynamic properties of the discrete-time PSO approach those of the continuous-time PSO when the time step approaches 0.

While theoretical research on PSO has mainly focused on convergence, there may be other theoretical properties that are more relevant in the context of optimisation. The primary goal in optimisation is to obtain a solution of acceptable quality within reasonable time. Convergence may be neither sufficient, nor necessary to reach this goal. In particular, convergence is insufficient when stagnation occurs at non-optimal points in the solution space. Furthermore, stagnation is not necessary when a solution of acceptable quality has been found.

As an alternative measure, we suggest to consider for arbitrarily small $\varepsilon > 0$ the expected time until the algorithm for the first time obtains a search point x for which $|f(x) - f(x^*)| < \varepsilon$, where $f(x^*)$ is the function value of an optimal search

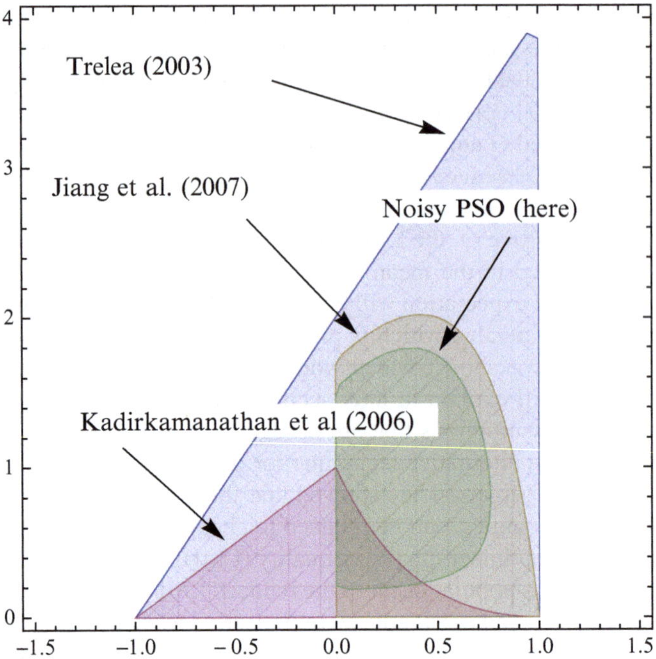

Fig. 1: Comparison of convergence regions. Noisy PSO indicates when the precondition $f(1) > 1/3$ of Theorem 2 holds. (x-axis: ω, y-axis: $\varphi = \varphi_1 = \varphi_2$)

point, where time is measured in the number of evaluations of the objective function. We call this the *expected first hitting time* (FHT) with respect to ε. As a first condition, it is desirable to have finite expected FHT for any constant $\varepsilon > 0$. Informally, this means that the algorithm will eventually find a solution of acceptable quality. Secondly, it is desirable that the growth of the expected FHT is upper bounded by a polynomial in $1/\varepsilon$ and the number of dimensions n of the problem. Informally, this means that the algorithm will not only find a solution of acceptable quality, but will do so within reasonable time.

Some work has been done in this direction. Sudholt and Witt [11] studied the runtime of the Binary PSO, i.e. in a discrete search space. Witt [15] considered the Guaranteed Convergence PSO (GCPSO) with one particle on the SPHERE function, showing that if started in unit distance to the optimum, then after $O(n\log(1/\varepsilon))$ iterations, the algorithm has reached the ε-ball around the optimum with overwhelmingly high probability. The GCPSO avoids stagnation by resetting the global best particle to a randomly sampled point around the best found position. The behaviour of the one-particle GCPSO therefore resembles the behaviour of a $(1+1)$ ES, and the velocity term does not come into play. In fact, the analysis has some similarities with the analysis by Jägersküpper [6].

The objectives of this paper are three-fold. Firstly, in Sect. 3, we show that the expected first hitting time of a basic PSO is infinite, even on the simple SPHERE function. Secondly, in Sect. 4, we point out situations where the basic PSO does not converge in mean square to the global best particle (which needs not be a global optimum), despite having parameters in the convergence region. We discuss what extra conditions are needed to ensure mean square convergence. Finally, in Sect. 5, we consider a Noisy PSO which we prove to have finite expected FHT on the one-dimensional SPHERE function. Our results also hold for any strictly increasing transformation of this function because the PSO is a comparison-based algorithm.

2 Preliminaries

In the following, we consider minimisation of functions. A basic PSO with swarm size m optimising an n-dimensional function $f : \mathbb{R}^n \to \mathbb{R}$ is defined below. This PSO definition is well-accepted, and called Standard PSO by Jiang et al. [7]. The position and velocity of particle $i \in [m]$ at time $t \geq 0$ are represented by the pair of vectors $X_t^{(i)} = (X_{t,1}^{(i)}, \ldots, X_{t,n}^{(i)})$ and $V_t^{(i)} = (V_{t,1}^{(i)}, \ldots, V_{t,n}^{(i)})$. The parameter $\alpha > 0$ bounds the initial positions and velocities.

Algorithm 1 Basic PSO

---

for each particle $i \in [m]$, and dimension $j \in [n]$ **do**
  $X_{0,j}^{(i)}, V_{0,j}^{(i)} \sim \text{Unif}[-\alpha, \alpha]$ $\;P_{0,j}^{(i)} = X_{0,j}^{(i)}$
$G_0 = \arg\min\{f(P_0^{(1)}), \ldots, f(P_0^{(m)})\}$
for $t = 0, 1, \ldots$ until termination condition satisfied **do**
  for each particle $i \in [m]$, and dimension $j \in [n]$ **do**

$$V_{t+1,j}^{(i)} = \omega V_{t,j}^{(i)} + \varphi_1 R_{t,j}^{(i)} \left(P_{t,j}^{(i)} - X_{t,j}^{(i)}\right) + \varphi_2 S_{t,j}^{(i)} \left(G_{t,j} - X_{t,j}^{(i)}\right), \tag{5}$$

$$X_{t+1,j}^{(i)} = X_{t,j}^{(i)} + V_{t+1,j}^{(i)}, \quad \text{where} \quad R_{t,j}^{(i)}, S_{t,j}^{(i)} \sim \text{Unif}[0,1]. \tag{6}$$

$$P_{t+1}^{(i)} = \arg\min\{f(X_t^{(i)}), f(P_t^{(i)})\} \quad \text{and}$$
$$G_{t+1} = \arg\min\{f(P_t^{(1)}), \ldots, f(P_t^{(m)})\}.$$

Assume that a function $f : \mathbb{R}^n \to \mathbb{R}$ has at least one global minimum x^*. Then for a given $\varepsilon > 0$, the *first hitting time* (FHT) of the PSO on function f is defined as the number of times the function f is evaluated until the swarm for the first time contains a particle x for which $|f(x) - f(x^*)| < \varepsilon$. We assume that the PSO is implemented such that the function f is evaluated no more than m times per time step t. As an example function, we consider the SPHERE problem, which for all $x \in \mathbb{R}^n$ is defined as $\text{SPHERE}(x) := \|x\|^2$, where $\|\cdot\|$ denotes the Euclidian norm. This is a well-accepted benchmark problem in convergence analyses and frequently serves as a starting point for theoretical analyses.

3 Stagnation

Particle convergence does not necessarily occur in local optima. There are well-known configurations, e.g. with zero velocities, which lead to stagnation [13]. However, it is not obvious for which initial configurations and parameter settings the basic PSO will stagnate outside local optima. Here, it is shown that stagnation occurs already with one dimension for a broad range of initial parameters. It follows that the expected first hitting time of the basic PSO can be infinite.

As a first example of stagnation, we consider the basic PSO with swarm size one on the SPHERE problem. Note that it is helpful to first study the PSO with swarm size one before analysing the behaviour of the PSO with larger swarm sizes. This is similar to the theory of evolutionary algorithms (EAs), where it is common to initiate runtime analyses on the simple $(1+1)$ EA with population size one, before proceeding to more complex EAs.

Proposition 1. *The basic PSO with inertia factor $\omega < 1$ and one particle $(m = 1)$ has infinite expected FHT on* SPHERE *$(n = 1)$.*

Proof. We say that the *bad initialisation event* has occurred if the initial position and velocity satisfy $X_0 > \varepsilon\alpha$ and $(\varepsilon\alpha - X_0) \cdot (1-\omega) < V_0 < 0$. This event occurs with positive probability. We claim that if the event occurs, then in any iteration $t > 0$, $V_t = V_0\omega^{t-1}$, and $X_t = X_0 + V_0 \sum_{i=1}^{t-1} \omega^i$. If the claim holds, then for all $t \geq 0$, it holds that $X_t < X_{t-1}$ and $G_t = X_t$. Therefore,

$$G_t > X_0 + V_0 \sum_{i=1}^{\infty} \omega^t > X_0 + \varepsilon\alpha - X_0 = \varepsilon\alpha.$$

and the proposition follows. Note that since $G_t = X_t$ for each $t > 0$, the velocity reduces to $V_t = \omega V_{t-1}$.

The claim is proved by induction on t. The base case $t = 2$ clearly holds, because $V_1 = V_0\omega$ and $X_1 = X_0 + V_0\omega$. Assume the claim holds for all iterations smaller than t. By induction, it holds that $V_t = \omega V_{t-1} = \omega^{t-1}V_0$. Therefore, by the induction hypothesis,

$$X_t = X_{t-1} + V_t = X_0 + V_0 \sum_{i=1}^{t-2} \omega^i + V_0\omega^{t-1} = X_0 + V_0 \sum_{i=1}^{t-1} \omega^i.$$

The claim now holds for all $t > 0$. The expected FHT conditional on the bad initialisation event is therefore infinite. However, the bad initialisation event occurs with positive probability, so the unconditional expected FHT is infinite by the law of total probability. □

We prove that the stagnation on SPHERE illustrated in Proposition 1 is not an artefact of a trivial swarm size of 1. In the following theorem, we prove stagnation for a swarm of size 2 and think that the ideas can be generalised to bigger swarm sizes. We allow any initialisation of the two particles that are sufficiently far away from the optimum. It is assumed that both velocities are non-positive in the initialisation step, which event occurs with constant probability for uniformly drawn velocities.

Theorem 1. *Consider the basic PSO with two particles on the one-dimensional* SPHERE. *If* $\omega < 1$, $1 < \varphi_2 < 2$, $V_0^{(1)}, V_0^{(2)} \leq 0$, $\kappa < 1$ *where*

$$\kappa := \frac{\varphi_2^2 - 2\varphi_2 + 2 + 2\omega\varphi_2}{4\varphi_2} + \frac{\sqrt{(\varphi_2^2 - 2\varphi_2 + 2\omega\varphi_2 + 2)(\varphi_2^2 + 6\varphi_2 + 2\omega\varphi_2 + 2)}}{4\varphi_2},$$

$$\textit{and} \quad X_0^{(1)}, X_0^{(2)} > 2\varepsilon + 2\varphi_2 \cdot \left(\frac{|X_0^{(2)} - X_0^{(1)}| + |V_0^{(1)}| + |V_0^{(2)}|}{(1-\omega)(1-\kappa)} \right)$$

all hold together, then the expected FHT for the ε*-ball around the optimum is infinite.*

The conditions are fulfilled, e.g., if $\varphi_2 = 1.5$, $\omega = 0.07$, $\varepsilon = 0.5$, $V_0^{(1)} = V_0^{(2)} = -1$, $X_0^{(1)} = 184$, and $X_0^{(2)} = 185$. For a proof, we note that the assumed initialisation with positive particle positions, negative velocities and α sufficiently large makes the sequences $X_t^{(i)}$, $i = 1, 2$, non-increasing provided no negative values are reached. Furthermore, the update equation for the velocities will then consist of three random non-positive terms, which means that velocities remain negative. In Lemma 2, we focus on the distance $D_t := X_t^{(2)} - X_t^{(1)}$ of the particles and show that its expectation converges absolutely to zero. The proof of this lemma makes use of Lemma 1, which gives a closed-form solution to a generalisation of the Fibonacci-sequence. In another lemma, we consider the absolute velocities over time and show that the series formed by these also converges in expectation. The proof of the theorem will be completed by applications of Markov's inequality.

Lemma 1. *For any real* $c > 0$*, there exists two reals* A *and* B *such that the difference equation* $a_n = c(a_{n-1} + a_{n-2}), n \geq 1$*, has the solution* $a_n = \alpha^n A + \beta^n B$*, where*

$$\alpha = \frac{c - \sqrt{c(4+c)}}{2}, \quad \textit{and} \quad \beta = \frac{c + \sqrt{c(4+c)}}{2}.$$

Proof. The proof is by induction over n. The lemma can always be satisfied for $n = 1$ and $n = 2$ by choosing appropriate A and B. Hence, assume that the lemma holds for all $i < n$ for some A and B. Note that

$$c(\alpha + 1) = \frac{c^2 - c\sqrt{c(4+c)} + 2c}{2} = \alpha^2, \quad \text{and}$$

$$c(\beta + 1) = \frac{c^2 + c\sqrt{c(4+c)} + 2c}{2} = \beta^2$$

It therefore follows by the induction hypothesis that

$$\begin{aligned} a_n &= c(a_{n-1}+a_{n-2}) \\ &= c(\alpha^{n-1}A+\beta^{n-1}B+\alpha^{n-2}A+\beta^{n-2}B) \\ &= c\alpha^{n-2}(\alpha+1)A+c\beta^{n-2}(\beta+1)B \\ &= \alpha^n A+\beta^n B. \end{aligned}$$

□

Lemma 2. *Given $t \geq 1$, suppose that for all $s \leq t$ it holds that $X_s^{(1)}, X_s^{(2)} \geq 0$ and $V_s^{(1)}, V_s^{(2)} \leq 0$. Then $E(|D_t|) \leq \kappa^t(2|D_0| + V_0^{(1)} - V_0^{(2)})$.*

Proof. The proof is mainly based on an inspection of the update equation of PSO. The aim is to obtain a recurrence for $E(|D_t|)$, where we have to distinguish between two cases. We abbreviate $\varphi = \varphi_2$ and $S = S_t^{(\cdot)}$ in the following.

If $X_t^{(1)} \leq X_t^{(2)}$, then $G_t = X_t^{(1)}$ and the update equations are

$$\begin{aligned} V_{t+1}^{(1)} &= \omega V_t^{(1)} \\ V_{t+1}^{(2)} &= \omega V_t^{(2)} + S\varphi(X_t^{(1)} - X_t^{(2)}) = \omega V_t^{(2)} - S\varphi D_t, \end{aligned}$$

which means $D_{t+1} = X_{t+1}^{(2)} - X_{t+1}^{(1)} = X_t^{(2)} - X_t^{(1)} - S\varphi D_t + \omega(V_t^{(2)} - V_t^{(1)})$. Since $V_{t+1}^{(i)} = X_{t+1}^{(i)} - X_t^{(i)}$, for $i = 1,2$, we obtain $V_t^{(2)} - V_t^{(1)} = D_t - D_{t-1}$, for $t \geq 0$, where we define $D_{-1} = D_0 - V_0^{(2)} + V_0^{(1)}$ to make the equation apply also for $t = 0$. Together, this gives us

$$D_{t+1} = D_t - S\varphi D_t + \omega(D_t - D_{t-1}). \tag{7}$$

If $X_t^{(1)} \geq X_t^{(2)}$, then the update equations are

$$V_{t+1}^{(1)} = \omega V_t^{(1)} + S\varphi(X_t^{(2)} - X_t^{(1)}) = \omega V_t^{(1)} + S\varphi D_t$$

and $V_{t+1}^{(2)} = \omega V_t^{(2)}$, which again results in Eq. (7) and finishes the case analysis. Taking absolute values on both sides of Eq. (7) and applying the triangle inequality to the right-hand side, we get

$$|D_{t+1}| \leq |(1 - S\varphi + \omega)||D_t| + |\omega||D_{t-1}|.$$

After taking the expectation and noting that $\omega > 0$, we have

$$E(|D_{t+1}| \mid D_t, D_{t-1}) \leq (E(|1 - S\varphi|) + \omega)|D_t| + \omega|D_{t-1}|.$$

The so-called law of total expectation states that $E(E(X \mid Y)) = E(X)$ for arbitrary random variables X and Y. As the conditional probability above is linear in both

$|D_{t-1}|$ and $|D_t|$, applying the law of total expectation and linearity of expectation twice yields

$$E(|D_{t+1}|) \leq (E(|1-S\varphi|)+\omega)E(|D_t|)+\omega E(|D_{t-1}|).$$

We are left with an estimate for $E(|1-S\varphi|)$. By the law of total probability,

$$\begin{aligned} E(|1-S\varphi|) = {} & E(1-S\varphi \mid 1-S\varphi \geq 0)\Pr(1-S\varphi \geq 0) \\ & + E(S\varphi - 1 \mid 1-S\varphi \leq 0)\Pr(1-S\varphi \leq 0). \end{aligned}$$

Since S is uniformly distributed and $\varphi > 1$, the first conditional expectation is $1/2$, while the second conditional expectation is $(\varphi-1)/2$. The probabilities for the conditions to occur are $1/\varphi$ and $1-1/\varphi$, respectively. This results in

$$E(|1-S\varphi|) = \frac{1}{2\varphi} + \left(1-\frac{1}{\varphi}\right)\cdot\frac{\varphi-1}{2} = \frac{\varphi^2-2\varphi+2}{2\varphi},$$

which finally gives us the following recurrence on $E(|D_t|)$:

$$E(|D_{t+1}|) \leq \left(\frac{\varphi^2-2\varphi+2}{2\varphi}+\omega\right)(E(|D_t|)+E(|D_{t-1}|)).$$

Introducing $D_t^* := E(|D_t|)$ and using $\lambda = \frac{\varphi^2-2\varphi+2}{\phi} + 2\omega$, we have in more compact form that

$$D_{t+1}^* \leq \frac{\lambda}{2}(D_t^* + D_{t-1}^*)$$

for $t \geq 0$. Solving this recursion (noting that all terms are positive) using Lemma 1 yields for $t \geq 1$ that

$$D_t^* \leq \left(\frac{\lambda - \sqrt{8\lambda+\lambda^2}}{4}\right)^t D_{-1}^* + \left(\frac{\lambda + \sqrt{8\lambda+\lambda^2}}{4}\right)^t D_0^*.$$

Note that $\kappa = \frac{\lambda+\sqrt{8\lambda+\lambda^2}}{4} < 1$ if and only if $\lambda < 1$. Furthermore, the factor in front of D_{-1}^* has clearly smaller absolute value than κ. We obtain

$$D_t^* \leq \kappa^t(D_{-1}^* + D_0^*) \leq \kappa^t(2D_0^* + V_0^{(1)} - V_0^{(2)}),$$

which we wanted to show. □

The following lemma uses the previous bound on $E(|D_t|)$ to show that the expected sum of velocities converges absolutely over time. This means that the maximum achievable progress is bounded in expectation. As an example, when choosing $\varphi_2 = 1.5$ and $\omega = 0.07$ in the following, we obtain a value of about $191(|D_0| + |V_0^{(1)}| + |V_0^{(2)}|)$ for this bound.

Lemma 3. *Suppose the prerequisites of Lemma 2 apply. Then for $i = 1,2$ it holds that*

$$\sum_{t=0}^{\infty} E(|V_t^{(i)}|) \leq \left(\frac{2\varphi_2}{(1-\omega)(1-\kappa)}\right)(|D_0| + |V_0^{(1)}| + |V_0^{(2)}|).$$

Proof. For notational convenience, we drop the upper index and implicitly show the following for both $i = 1$ and $i = 2$. According to the update equation of PSO, we have $|V_{t+1}| \leq w|V_t| + \varphi|D_t|$ for $t \geq 0$, using $\varphi := \varphi_2$. Resolving the recurrence yields for $t \geq 1$ that

$$|V_t| \leq \omega^t|V_0| + \varphi \sum_{s=0}^{t-1} \omega^s |D_{t-1-s}| = \omega^t|V_0| + \varphi \sum_{s=0}^{t-1} \omega^{t-1-s}|D_s|.$$

Hence,

$$\begin{aligned}\sum_{t=1}^{\infty}|V_t| &\leq \sum_{t=1}^{\infty}\left(\omega^t|V_0| + \varphi\sum_{s=0}^{t-1}\omega^{t-1-s}|D_s|\right)\\ &\leq \frac{|V_0|}{1-\omega} + \varphi\left(\sum_{t=0}^{\infty}\left(\sum_{s\geq t}\omega^s\right)|D_t|\right) \leq \frac{\varphi}{1-\omega}\left(|V_0| + \sum_{t=0}^{\infty}|D_t|\right),\end{aligned}$$

since $0 < \omega < 1$ and $\varphi > 1$. Using the linearity of expectation,

$$\sum_{t=0}^{\infty} E(|V_t|) \leq \frac{\varphi}{1-\omega}\left(|V_0| + \sum_{t=0}^{\infty} E(|D_t|)\right).$$

By Lemma 2,

$$E(|D_t|) \leq \kappa^t(2|D_0| + V_0^{(1)} - V_0^{(2)}).$$

Hence, the series over the $E(|D_t|)$ converges according to

$$\sum_{t=0}^{\infty} E(|D_t|) \leq \frac{1}{1-\kappa}(2|D_0| + V_0^{(1)} - V_0^{(2)}),$$

which yields

$$\begin{aligned}\sum_{t=0}^{\infty} E(|V_t|) &\leq \frac{\varphi}{1-\omega}(V_0^{(1)} + V_0^{(2)}) + \frac{\varphi}{1-\omega}\frac{1}{1-\kappa}(2|D_0| + V_0^{(1)} - V_0^{(2)})\\ &\leq \left(\frac{\varphi}{(1-\omega)(1-\kappa)}\right)\cdot(2|D_0| + 2|V_0^{(1)}| + 2|V_0^{(2)}|),\end{aligned}$$

where we have used $\kappa < 1$. □

We are ready to prove Theorem 1.

Proof of Theorem 1 Throughout this proof, we suppose the prerequisites from Lemma 2 to hold. As we will show, these hold for an infinite number of steps with constant probability.

For any finite t, Lemma 3 and linearity of expectation yield for $i = 1,2$ that

$$E\left(\sum_{s=0}^{t}|V_s^{(i)}|\right) \leq \left(\frac{2\varphi_2}{(1-\omega)(1-\kappa)}\right)\cdot(|D_0|+|V_0^{(1)}|+|V_0^{(2)}|),$$

which by Markov's inequality means that the event

$$\sum_{s=0}^{t}|V_s^{(i)}| \leq \varepsilon + \left(\frac{2\varphi_2}{(1-\omega)(1-\kappa)}\right)\cdot(|D_0|+|V_0^{(1)}|+|V_0^{(2)}|)$$

occurs with a positive probability that does not depend on t. Given the assumed initial values of $X_0^{(i)}$, the ε-ball around the optimum is not reached if the event occurs. Hence, there is a probability $p^* > 0$ such that for any finite number of steps t, the probability of not hitting the ε-ball within t steps is at least p^*. Consequently, the expected first hitting time of the ε-ball is infinite. □

4 Mean Square Convergence

As mentioned in the introduction, there exist several convergence analyses using different techniques that take into account the stochastic effects of the algorithm. The analysis by Jiang et al. [7] is perhaps the one where the proof of mean square convergence follows most directly from the definition. They consider the basic PSO and prove the following statement (Theorem 5 in their paper):

Statement 1. *Given* $\omega, \varphi_1, \varphi_2 \geq 0$, *if* $0 \leq \omega < 1$, $\varphi_1 + \varphi_2 > 0$, *and* $0 < -(\varphi_1 + \varphi_2)\omega^2 + \left(\frac{1}{6}\varphi_1^2 + \frac{1}{6}\varphi_2^2 + \frac{1}{2}\varphi_1\varphi_2\right)\omega + \varphi_1 + \varphi_2 - \frac{1}{3}\varphi_1^2 - \frac{1}{3}\varphi_2^2 - \frac{1}{2}\varphi_1\varphi_2 < \frac{\varphi_2^2(1+\omega)}{6}$ *are all satisfied together, the basic particle swarm system determined by parameter tuple* $\{\omega, c_1, c_2\}$ *will converge in mean square to G.*

This statement is claimed to hold for any fitness function and for any initial swarm configuration. However, as acknowledged by the corresponding author (Jiang M., 2011), there is an error in the proof of the above statement, which is actually wrong without additional assumptions.

Intuitively Statement 1 makes sense for well-behaved, continuous functions like SPHERE. However, in retrospect, it is not too difficult to set up artificial fitness functions and swarm configurations where the statement is wrong: Let us consider the one-dimensional function $f : \mathbb{R} \to \mathbb{R}$ defined by

$$f(x) = \begin{cases} 0 & \text{if } x = 0, \\ 1 & \text{if } x = 1, \\ 2 & \text{otherwise,} \end{cases}$$

which is to be minimised.

Assume a swarm of two particles, where the first one has position 0, which is then its local best and the global best. Furthermore, assume velocity 0 for this particle, i.e., it has stagnated. Formally, $X_0^{(1)} = P_0^{(1)} = G_0 = V_0^{(1)} = 0$. Now let us say the second particle has current and local best position 1 and velocity 0, formally $X_0^{(2)} = P_0^{(2)} = 1$ and $V_0^{(2)} = 0$. This particle will now be attracted by a weighted combination of local and global best, e.g. the point 0.5 if both learning rates are the same. The problem is that the particle's local best almost surely will never be updated again since the probability of sampling either local best or global best is 0 if the sampling distribution is uniform on an interval of positive volume or is the sum of two such distributions, as it is defined in the basic PSO. The sampling distribution might be deterministic because both $P_t^{(i)} - X_t^{(i)}$ and $G_t - X_t^{(i)}$ might be 0, but then the progress corresponds to the last velocity value, which again was either obtained according to the sum of two uniform distributions or was already 0. The error in the analysis is hidden in the proof of Theorem 4 in [7], where $\Pr(X_t = G) > 0$ is concluded even though G might be in a null set. Nevertheless, important parts of the preceding analysis can be saved and a theorem on convergence can be proved under additional assumptions on the fitness function. In the following, we describe the main steps in the convergence analysis by Jiang et al. [7]. A key idea in [7] is to consider a one-dimensional algorithm and an arbitrary particle, assuming that the local best for this particle and global best do not change. Then a recurrence relation is obtained as follows:

$$X_{t+1} = (1 + \omega - (\varphi_1 R_t + \varphi_2 S_t))X_t - \omega X_{t-1} + \varphi_1 R_t P + \varphi_2 S_t G,$$

where we dropped the index denoting the arbitrary particle we have chosen, and the time index for local and global best. The authors proceed by deriving sufficient conditions for the sequence of expectations $E(X_t)$, $t \geq 1$, to converge (Theorem 1 in their paper).

Lemma 4. *Given $\omega, \varphi_1, \varphi_2 \geq 0$, if and only if $0 \leq \omega < 1$ and $0 < c_1 + c_2 < 4(1+\omega)$, the iterative process $E(X_t)$ is guaranteed to converge to $(\varphi_1 P + \varphi_2 G)/(\varphi_1 + \varphi_2)$.*

Even though a process converges in expectation, its variance might diverge, which intuitively means that it becomes more and more unlikely to observe the actual process in the vicinity of the expected value. Another major achievement by Jiang et al. [7] is to study the variances $\mathrm{Var}(X_t)$ of the still one-dimensional process. By a clever analysis of a recurrence of order 3, they obtain the following lemma (Theorem 3 in their paper).

Lemma 5. *Given $\omega, \varphi_1, \varphi_2 \geq 0$, if and only if $0 \leq \omega < 1$, $\varphi_1 + \varphi_2 \geq 0$ and $f(1) > 0$ are all satisfied together, iterative process $\mathrm{Var}(X_t)$ is guaranteed to converge to $\frac{1}{6}(\varphi_1 \varphi_2/(\varphi_1 + \varphi_2))^2 + (G-P)^2(1+\omega)/f(1)$, where*

$$f(1) = -(\varphi_1 + \varphi_2)\omega^2 + (\frac{1}{6}\varphi_2^2 + \frac{1}{2}\varphi_1^2\varphi_2^2)\omega + \varphi_1 + \varphi_2 - \frac{1}{3}\varphi_1^2 - \frac{1}{3}\varphi_2^2 - \frac{1}{2}\varphi_1\varphi_2.$$

Lemma 5 means that the variance is proportional to $(P-G)^2$. However, in contrast to what Jiang et al. [7] would like to achieve in their Theorem 4, we do not see how to prove that the variance approaches 0 for every particle. Clearly, this happens for the global best particle under the assumption that no further improvements of the global best are found. We do not follow this approach further since we are interested in PSO variants that converge to a local optimum.

5 Noisy PSO

The purpose of this section is to consider a variant of the basic PSO that includes a noise term. This PSO, which we call the Noisy PSO, is defined as in Algorithm 1, except that Eq. (5) is replaced by the velocity equation

$$V_{t+1,j}^{(i)} = \omega V_{t,j}^{(i)} + \varphi_1 R_{t,j}^{(i)}(P_{t,j}^{(i)} - X_{t,j}^{(i)}) + \varphi_2 S_{t,j}^{(i)}(G_{t,j} - X_{t,j}^{(i)}) + \Delta_{t,j}^{(i)},$$

where the extra noise term $\Delta_{t,j}^{(i)}$ has uniform distribution on the interval $[-\delta/2, \delta/2]$. Note that our analysis seems to apply also when the uniform distribution is replaced by a Gaussian one with the same expectation and variance. The constant parameter $\delta > 0$ controls the noise level in the algorithm. Due to the uniformly distributed noise term, it is immediate that the variance of each particle is always at least $\delta^2/12$. Therefore, the Noisy PSO does not enjoy the mean square convergence property of the basic PSO. In return, the Noisy PSO does not suffer the stagnation problems discussed in Sect. 1, and finite expected first hitting times can in some cases be guaranteed. The noisy PSO uses similar measures to avoid stagnation as the GCPSO mentioned in the introduction. However, our approach is simpler and treats all particles in the same way. On the other hand, the GCPSO relies on a specific update scheme for the global best particle.

Our main result considers the simplified case of a one-dimensional function but takes into account the whole particle swarm. For simplicity, we only consider the half-open positive interval by defining

$$\text{SPHERE}^+(x) := \begin{cases} \text{SPHERE}(x) & \text{if } x \geq 0 \\ \infty & \text{otherwise,} \end{cases}$$

which has to be minimised, and assume that at least one particle is initialised in the positive region. This event happens with positive probability for a standardised initialisation scheme. It seems that our analyses can be adapted to the standard SPHERE (and order-preserving transformations thereof), but changes of sign complicate the analysis considerably. Note that the analyses of stagnation in Sect. 3 only consider positive particle positions and thus apply to SPHERE^+ as well.

Theorem 2. *Consider the Noisy PSO on the* SPHERE^+ *function and assume* $G_0 \geq 0$. *If* $\delta \leq \varepsilon, f(1) > 1/3$ *(for the* f *from Lemma 5) and the assumptions from Lemmas 4* *and 5 hold (which are the same as in Theorems 3 and 4 below, respectively), then the expected first hitting time for the interval* $[0, \varepsilon]$ *is finite.*

The proof of this theorem relies heavily on the convergence analysis by Jiang et al. [7]. We will adapt their results to the Noisy PSO. Recall that the only difference between the two algorithms is the addition of $\Delta^{(i)}$ in the update equation for the particle position. It is important to note that $\Delta^{(i)}$ is drawn from $[-\delta/2, \delta/2]$ (considering one dimension) fully independently for every particle and time step. As mentioned above, Jiang et al. [7] consider a one-dimensional algorithm and an arbitrary particle, assuming that the local best for this particle and global best do not change. Then a recurrence relation is obtained by manipulating the update equations. Taking this approach for the Noisy PSO yields:

$$X_{t+1} = (1+\omega-(\varphi_1 R_t+\varphi_2 S_t))X_t-\omega X_{t-1}+\varphi_1 R_t P+\varphi_2 S_t G+\Delta_t,$$

where we dropped the index for the dimension, the index denoting the arbitrary particle we have chosen and the time index for local and global best. This is the same recurrence relation as in [7] except for the addition of Δ_t. The authors proceed by deriving sufficient conditions for the sequence of expectations $E(X_t)$, $t \geq 1$, to converge. Since $E(\Delta_t) = 0$, the recurrence relation for the expectations is *exactly* the same as with the basic PSO and the following theorem can be taken over.

Theorem 3. *Given* $\omega, \varphi_1, \varphi_2 \geq 0$, *if and only if* $0 \leq \omega < 1$ *and* $0 < \varphi_1 + \varphi_2 < 4(1+\omega)$, *the iterative process* $E(X_t)$ *is guaranteed to converge to* $(\varphi_1 P + \varphi_2 G)/(\varphi_1 + \varphi_2)$.

The next step is to study the variances $\mathrm{Var}(X_t)$ of the one-dimensional process. Obviously, modifications of the original analysis in [7] become necessary here. To account for the addition of Δ_t, we replace Eq. (11) in the paper[1] by

$$Y_{t+1} = (\psi - R_t)Y_t - \omega Y_{t-1} + Q'_t,$$

where $Q'_t := Q_t + \Delta_t$ and Q_t is the original Q_t from the paper. Regarding the quantities involving Q'_t in the following, we observe that $E(Q'_t) = E(Q_t) = 0$ and

$$\begin{aligned}\mathrm{Var}(Q'_t) &= E((Q'_t)^2) = E((Q_t+\Delta_t)(Q_t+\Delta_t)) \\ &= E(Q_t^2 + 2\Delta_t Q_t + \Delta_t^2) = E(Q_t^2) + E(\Delta_t^2),\end{aligned}$$

where we used that Δ_t is drawn independently of other random variables. Finally, we get

$$E(R_t Q'_t) = E(R_t(Q_t+\Delta_t))) = E(R_t Q_t),$$

which means that all following calculations in Section 3.2 in [7] may use the same values for the variables R and T as before. Only the variable Q increases by $E(\Delta_t^2)$. Recall that $\Delta_t \sim U[-\delta/2, \delta/2]$ for constant $\delta > 0$. We obtain $E(\Delta_t^2) = \delta^2/12$. Now

[1] When referring to the analysis by Jiang et al. [7], R_t does not mean the random factor in the cognitive component, but should be understood as defined in their paper.

the iteration equation (17) for $\mathrm{Var}(X_t)$ can be taken over with Q increased by $\delta^2/12$. The characteristic equation (18) remains unchanged and Theorem 2 applies in the same way as before. Theorem 3 in [7] is updated in the following way and proved as before, except for plugging in the updated value of Q.

Theorem 4. *Given $\omega, \varphi_1, \varphi_2 \geq 0$, if and only if $0 \leq \omega < 1$, $\varphi_1 + \varphi_2 \geq 0$ and $f(1) > 0$ are all satisfied together, iterative process $\mathrm{Var}(X_t)$ is guaranteed to converge to $(\frac{1}{6}(\varphi_1\varphi_2/(\varphi_1+\varphi_2))^2(G-P)^2(1+\omega)) + \delta^2/12)/f(1)$, where*

$$f(1) = -(\varphi_1+\varphi_2)\omega^2 + \left(\frac{1}{6}\varphi_2^2 + \frac{1}{2}\varphi_1^2\varphi_2^2\right)\omega + \varphi_1 + \varphi_2 - \frac{1}{3}\varphi_1^2 - \frac{1}{3}\varphi_2^2 - \frac{1}{2}\varphi_1\varphi_2.$$

As a consequence from the preceding lemma, the variance remains positive even for the particle i that satisfies $P^{(i)} = G$. Under simplifying assumptions, we show that this particle allows the system to approach the optimum. Later, we will show how to drop the assumption.

Lemma 6. *Assume that $f(1) > 1/3$, that the global and local bests are never updated, and that the conditions in Theorems 3 and 4 hold. Then for all sufficiently small $\varepsilon' > 0$, there exists a $t_0 > 0$ such that*

$$\forall t \geq t_0 \quad \Pr\left[G - \delta \leq X_t \leq G - \delta/100 + \varepsilon'\right] \geq 3/100000,$$

where $X_t \in \mathbb{R}$ is the position of the particle in iteration t for which the local best position equals the global best position.

Proof. We assume that $G > \varepsilon$, otherwise there is nothing to show. Furthermore, $G - \delta$ cannot be negative since $\delta \leq \varepsilon$. We decompose the process by defining $Y_t = X_t - \Delta_t$. Our goal is to prove that it is unlikely that Y_t is much larger than G using Chebyshev's inequality. We therefore need to estimate the expectation and the variance of Y_t. From Theorem 3 and the fact that $P = G$ holds for the best particle,

$$\lim_{t\to\infty} E(Y_t) = \lim_{t\to\infty} E(X_t) - E(\Delta_t) = G. \tag{8}$$

To estimate the variance of Y_t, first recall that by Theorem 4, it holds that

$$\lim_{t\to\infty} \mathrm{Var}(X_t) = \delta^2/(12f(1)). \tag{9}$$

Due to the independence of the random variables Y_t and Δ_t, we have $\mathrm{Var}(X_t) = \mathrm{Var}(Y_t) + \mathrm{Var}(\Delta_t)$. The random variable Δ_t has variance $\delta^2/12$. The limit in Eq. (9) therefore implies that

$$\lim_{t\to\infty} \mathrm{Var}(Y_t) = \lim_{t\to\infty} \mathrm{Var}(X_t) - \mathrm{Var}(\Delta_t) = \sigma_Y^2, \tag{10}$$

where we have defined

$$\sigma_Y^2 := \delta^2(1-f(1))/(12f(1)) \leq \delta^2/6.$$

Combining Eqs. (8) and (10), yields

$$\lim_{t\to\infty} E(Y_t) + (6/5)\sqrt{\mathrm{Var}(Y_t)} = G + (6/5)\sigma_Y.$$

This limit implies that for any $\varepsilon' > 0$, there exists a $t_0 > 0$ such that

$$\begin{aligned}\forall t \geq t_0 \quad E(Y_t) + (6/5)\sqrt{\mathrm{Var}(Y_t)} &\leq G + (6/5)\sigma_Y + \varepsilon' \\ &\leq G + 0.4899\delta + \varepsilon',\end{aligned}$$

and analogously $E(Y_t) - (6/5)\sqrt{\mathrm{Var}(Y_t)} \geq G + 0.4899\delta + \varepsilon'$. By the inequality above, and by Chebyshev's inequality, it holds that

$$p := \Pr\left[|Y_t - G| \geq 0.4899\delta + \varepsilon'\right] \leq \Pr\left[|Y_t - E(Y_t)| \geq (6/5)\sqrt{\mathrm{Var}(Y_t)}\right] \leq 25/36.$$

Obviously, the larger Y_t is, the more restrictive the requirements on the outcome of Δ_t are. Hence, choosing t so large that $\varepsilon' \leq (1-0.4899)\delta$ holds, we get the desired result

$$\begin{aligned}\Pr\left[G - \delta \leq X_t < G - \delta/100 + \varepsilon'\right] &\geq \Pr\left[\Delta_t < -\delta(1/100 + 0.4899)\right] \cdot (1-p) \\ &\geq 3/100000.\end{aligned}$$

□

The previous lemma does not make any assumption on the objective function. With regard to SPHERE$^+$, it implies that the global best (assuming $G_0 > 0$) will be improved after some time almost surely. However, since the precondition is that the particle has not improved for a while, this is not yet sufficient to ensure finite hitting time to an ε-ball around the optimum. One might imagine that the global best position is constantly updated while its value converges to some value greater than 0.

A closer look into the proofs of Lemmas 3 and 4 and the underlying difference equations in [7] reveals that they also apply to every particle i where $(G_t - P_t^{(i)})^2$ converges to a fixed value. In fact, as we will show in Lemma 7, it holds that $(G_t - P_t^{(i)})^2$ almost surely is a null sequence for every positively initialised particle if certain assumptions on the parameters are met. Informally, this means that the personal best converges to the global best on SPHERE$^+$, which might be considered as a corrected version of the erroneous Statement 1 in [7].

Lemma 7. *Consider the basic PSO on* SPHERE$^+$. *If* $f(1) > \max\{\varphi_1^2, \varphi_2^2\}(1+\omega)/6$ *and the assumptions from Theorems 3 and 4 hold, then* $(G_t - P_t^{(i)})^2$ *is a null sequence for every particle i that satisfies* $P_0^{(i)} \geq 0$. *The statement also holds for the Noisy PSO if additionally* $f(1) > 1/3$ *is assumed.*

Proof. If there is no particle satisfying $P_0^{(i)} \geq 0$, nothing is to show. Otherwise, we have $G_0 \geq 0$. Pick an arbitrary particle i satisfying $P_0^{(i)} \geq 0$ and assume that

$(G_t - P_t^{(i)})^2$ is not a null sequence for this particle. Because of the special properties of one-dimensional SPHERE$^+$ and the conditions $G_0 \geq 0$ and $P_0^{(i)} \geq 0$, the sequences G_t and $P_t^{(i)}$ are monotone decreasing and bounded, hence they are convergent. Assume that $(G_t - P_t^{(i)})^2$ converges to some non-zero value. According to Theorems 3 and 4, $E(X_t)$ and $\mathrm{Var}(X_t)$ converge, more precisely it holds for the expectation that $\lim_{t\to\infty} G_t < \lim_{t\to\infty} E(X_t) < \lim_{t\to\infty} P_t^{(i)}$. In the case of the basic PSO, we obtain from Theorems 3 and 4 that

$$\lim_{t\to\infty} E(X_t) = (\varphi_1 \lim_{t\to\infty} P_t^{(i)} + \varphi_2 \lim_{t\to\infty} G_t)/(\varphi_1 + \varphi_2)$$

and

$$\lim_{t\to\infty} \mathrm{Var}(X_t) = \frac{1}{6}\frac{\varphi_1\varphi_2}{(\varphi_1+\varphi_2)^2}(\lim_{t\to\infty} G_t - \lim_{t\to\infty} P_t^{(i)})^2 \frac{(1+\omega)}{f(1)}.$$

If $f(1) > \varphi_2^2(1+\omega)/6$, we obtain

$$\lim_{t\to\infty} \mathrm{Var}(X_t) < (\varphi_1/(\varphi_1+\varphi_2))^2(\lim_{t\to\infty} G_t - \lim_{t\to\infty} P_t^{(i)})^2 = (\lim_{t\to\infty} G_t - \lim_{t\to\infty} E(X_t))^2,$$

and if $f(1) > \varphi_1^2(1+\omega)/6$ we obtain

$$\lim_{t\to\infty} \mathrm{Var}(X_t) < (\lim_{t\to\infty} P_t^{(i)} - \lim_{t\to\infty} E(X_t))^2.$$

If both inequalities apply, then the variance is smaller than the smallest of the two squared distances, and Chebyshev's inequality yields that

$$|X_t - \lim_{t\to\infty} E(X_t)| < \min\{|\lim_{t\to\infty} P_t^{(i)} - \lim_{t\to\infty} E(X_t)|, |\lim_{t\to\infty} P_t^{(i)} - \lim_{t\to\infty} E(X_t)|\},$$

implying $G_t < X_t < P_t^{(i)}$, will occur with positive probability for sufficiently large t (using the same methods as in the proof of Lemma 6 the errors become negligible if t is large enough). This leads to an improvement of $P_t^{(i)}$ by a positive amount and also $(P_t^{(i)} - G_t)^2$ will decrease by a positive amount. Note the lower bound on the size of the positive improvement does not change as time increases. As t approaches infinity, the improvement will happen almost surely.

In the case of the Noisy PSO, the argumentation is similar. However, since the limit of the variance increases by $(\delta^2/12)/f(1)$ according to Theorem 4, we will decompose the stochastic process in the same way as in the proof of Lemma 6 and combine the calculations that follow from $f(1) > 1/3$ with the considerations presented above for the basic PSO. For the variable Y_t, Chebyshev's inequality yields that

$$|Y_t - \lim_{t\to\infty} E(X_t)| < |\lim_{t\to\infty} P_t^{(i)} - \lim_{t\to\infty} E(X_t)| + 0.4899\delta$$

and

$$|Y_t - \lim_{t\to\infty} E(X_t)| < |\lim_{t\to\infty} G_t - \lim_{t\to\infty} E(X_t)| + 0.4899\delta$$

both occur with positive probability. Hence, the support of $X_t = Y_t + \Delta$ is a superset of a subset of $[\lim_{t\to\infty} G_t, \lim_{t\to\infty} P_t^{(i)}]$ with positive measure. Therefore, an improvement by a certain positive amount has positive probability and will occur almost surely as t tends to infinity. □

Remark. The preconditions are satisfied for $\omega = 0.4$, $\varphi_1 = \varphi_2 = 1.5$, which is included in the convergence region of [7] (cf. Fig. 1). We can formulate the announced generalisation of Lemma 6.

Lemma 8. *Assume that $f(1) > 1/3$ and that the conditions in Theorems 3 and 4 hold. Consider the noisy PSO on* SPHERE$^+$, *pick a particle i satisfying $P_0^{(i)} \geq 0$ and denote $G = \lim_{t\to\infty} G_t$. Then for all sufficiently small $\varepsilon' > 0$, there exists a $t_0 > 0$ such that*

$$\forall t \geq t_0 \quad \Pr\left[G^* - \delta \leq X_t \leq G^* - \delta/100 + \varepsilon'\right] \geq 3/100000,$$

where X_t is the position of the considered particle in iteration t.

Proof. By Lemma 7 $(G_t - P_t^{(i)})^2$ converges to 0 for particle i. The stability analysis of the inhomogeneous difference equations for $E\left(X_t^{(i)}\right)$ and $\mathrm{Var}\left(X_t^{(i)}\right)$ does not change and we still get

$$\lim_{t\to\infty} E\left(X_t^{(i)}\right) = G \quad \text{and} \quad \lim_{t\to\infty} \mathrm{Var}\left(X_t^{(i)}\right) = \delta^2/(12f(1))$$

as in the proof of Lemma 6. The proof is completed as before. □

We are ready to prove the main result in this section.

Proof (of Theorem 2). Since it is monotonically decreasing and bounded, the sequence G_t has a limit G. If G is in the ε-ball around the origin, nothing is to show. Otherwise, we have $G > \varepsilon$ and according to Lemma 8, some point $X \in [G-\delta, G-\delta/100]$ will be sampled almost surely in finite time. After a finite number of such improvements, the ε-ball around the optimum has been reached. □

6 Experiments

The one-particle Standard PSO and Noisy PSO were implemented and applied to the two-dimensional SPHERE-problem with the parameter settings $\omega = 9/10$, $\varphi_1 = 1$, and $\alpha = 1$. The noise parameter in Noisy PSO was set to $\delta = 1/100$. In total 1,000 experiments were made for each algorithm. Each experiment lasted for 1,000 iterations. The distributions of the observed values of SPHERE are illustrated with box-and-whisker plots in Fig. 2 for every ten iterations. Each box-and-whisker plot shows the lower quartile, the median, and the upper quartile.

The tendency of Standard PSO to stagnate at non-optimal values is visible in the convergence of the distributions of SPHERE-values after 150 iterations. Note that

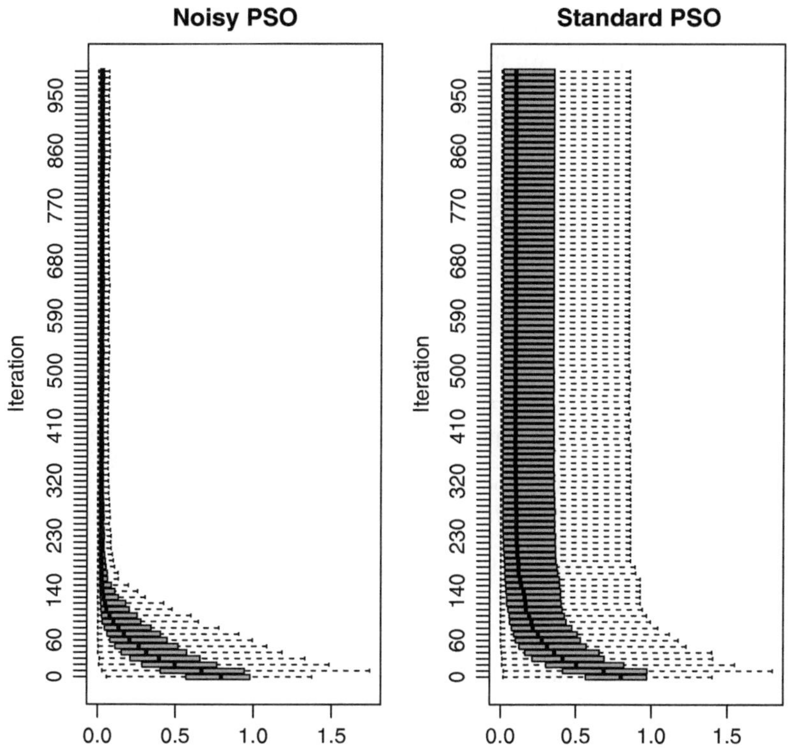

Fig. 2: Distribution of SPHERE-values as a function of iteration number after 1,000 experiments with Noisy PSO (*left*) and Standard PSO (*right*)

there is a large variance in the distribution of the point of stagnation. This does not contradict Statement 1, which only states that all swarm members will finally converge to the point of stagnation.

7 Conclusions

Much of the theoretical research on the particle swarm optimiser has focused on its convergence properties. In particular, conditions have been found which has been claimed to guarantee mean square convergence. We point out an error in the proof of this claim, showing that the mean square convergence property does not hold for all functions. Still, we think particle convergence is not always desirable, in particular when it occurs in non-optimal points in the search space. To better understand the PSO as an optimiser, we suggest to put more effort in understanding the expected first hitting time (FHT) of the algorithm to an arbitrarily small ε-ball around the optimum. We point out non-trivial configurations where the basic PSO has infinite expected FHT on even simple problems like the SPHERE function. As a remedy to this undesirable situation, we abandon convergence in mean square, and propose the Noisy PSO which has non-zero particle variance, but finite expected FHT on the one-dimensional SPHERE function.

Acknowledgements The authors thank Ming Yiang for helpful discussions. Supported by Deutsche Forschungsgemeinschaft (DFG) under grant no. WI 3552/1-1.

References

1. Boyd, S., El Ghaoui, L., Feron, E., Balakrishnan, V.: Linear matrix inequalities in system and control theory. In: Studies in Applied Mathematics, vol. 15. SIAM, Philadelphia (1994)
2. Brandstätter, B., Baumgartner, U.: Particle swarm optimization - mass-spring system analogon. IEEE Trans. Magn. **38**(2), 997–1000 (2002)
3. Clerc, M., Kennedy, J.: The particle swarm – explosion, stability, and convergence in a multidimensional complex space. IEEE Trans. Evol. Comput. **6**(1), 58–73 (2002)
4. Droste, S., Jansen, T., Wegener, I.: Upper and lower bounds for randomized search heuristics in black-box optimization. Theor. Comput. Syst. **39**(4), 525–544 (2006)
5. Fernández-Martínez, J.L., García-Gonzalo, E.: Stochastic stability analysis of the linear continuous and discrete PSO models. IEEE Trans. Evol. Comput. **15**(3), 405–423 (2011)
6. Jägersküpper, J.: Algorithmic analysis of a basic evolutionary algorithm for continuous optimization. Theor. Comput. Sci. **379**(3), 329–347 (2007)
7. Jiang, M., Luo, Y.P., Yang, S.Y.: Stochastic convergence analysis and parameter selection of the standard particle swarm optimization algorithm. Inform. Process. Lett. **102**(1), 8–16 (2007)
8. Kadirkamanathan, V., Selvarajah, K., Fleming, P.: Stability analysis of the particle dynamics in particle swarm optimizer. IEEE Trans. Evol. Comput. **10**(3), 245–255 (2006)
9. Kennedy, J., Eberhart, R.C.: Particle swarm optimization. In: Proceedings of the IEEE International Conference on Neural Networks, pp. 1942–1948. IEEE Press (1995)
10. Poli, R.: Dynamics and stability of the sampling distribution of particle swarm optimisers via moment analysis. J. Artif. Evol. Appl. **2008**, 15:1–15:10 (2008)
11. Sudholt, D., Witt, C.: Runtime analysis of a binary particle swarm optimizer. Theor. Comput. Sci. **411**(21), 2084–2100 (2010)
12. Trelea, I.C.: The particle swarm optimization algorithm: convergence analysis and parameter selection. Inform. Process. Lett. **85**(6), 317–325 (2003)
13. van den Bergh, F.: An analysis of particle swarm optimizers. Ph.D. thesis, Department of Computer Science, University of Pretoria, South Africa (2002)
14. Wakasa, Y., Tanaka, K., Nishimura, Y.: Control-theoretic analysis of exploitation and exploration of the PSO algorithm. In: 2010 IEEE International Symposium on Computer-Aided Control System Design (CACSD), pp. 1807–1812 (2010)
15. Witt, C.: Why standard particle swarm optimisers elude a theoretical runtime analysis. In: Proceedings of Foundations of Genetic Algorithms 10 (FOGA'09), pp. 13–20. ACM, New York, USA (2009)

Using Performance Profiles for the Analysis and Design of Benchmark Experiments

Helio J.C. Barbosa, Heder S. Bernardino, and André M.S. Barreto

Abstract It is common to manipulate a large amount of data generated in the benchmarking process when comparing metaheuristics. Performance profiles are analytical tools for the visualization and interpretation of these results. Here we comment on their explanatory power, discuss novel variants, introduce a multicriterion view of the performance comparison, and also define performance profiles for each test-problem in a given benchmark suite. In order to illustrate the potential of performance profiles for both algorithms and test-problems, we apply them to the results of an optimization competition so that new facts are pointed out and conclusions are drawn.

1 Introduction

Optimization problems appear naturally in many areas as one is often interested in minimizing or maximizing quantities (such as cost or profit, respectively) but also in other situations, such as system identification, which can frequently be formulated as the minimization of a conveniently defined objective function. In constrained optimization problems one is interested in the minimization (or maximization) of a given objective function $f(x)$, where $x \in R^n$ is the vector of

Helio J.C. Barbosa
Laboratório Nacional de Computação Científica (LNCC), Petrópolis, RJ, Brazil

Universidade Federal de Juiz de Fora (UFJF), Juiz de Fora, MG, Brazil
e-mail: hcbm@lncc.br

Heder S. Bernardino
Laboratório Nacional de Computação Científica (LNCC), Petrópolis, RJ, Brazil
e-mail: hedersb@lncc.br

André M.S. Barreto
School of Computer Science, McGill University, Montreal, QC, Canada
e-mail: amsb@cs.mcgill.ca

L. Di Gaspero et al. (eds.), *Advances in Metaheuristics*, Operations Research/ Computer Science Interfaces Series 53, DOI 10.1007/978-1-4614-6322-1_2,

design/decision variables, subject to inequality constraints $g_q(x) \leq 0$, $q = 1, 2, \ldots, \bar{q}$ as well as equality constraints $h_q(x) = 0$, $q = \bar{q}+1, \ldots, m$. Additionally, the variables are usually subject to bounds $x_i^L \leq x_i \leq x_i^U, i = 1, \ldots, n$. Although many practical problems can be straightforwardly cast in this form, in many complex real world situations an explicit mathematical expression for f, g_q, or h_q, as functions of the vector of decision variables x, may not be available: the constraints are in fact a complex implicit function of x, and checking for feasibility requires an expensive simulation. Furthermore, derivatives of the objective function and/or constraints with respect to the design variables may be undefined, noisy, expensive or unavailable. This so called "black-box" optimization context, where only "order zero" information is available, is perhaps the best suited setting for the application of stochastic, population-based, nature-inspired techniques.

Over the years, many techniques for tackling this class of problems have been developed, and the need to assess their relative performance is thus obvious. In some cases, a theoretical worst- and/or average case analysis can be made, but the difficulties involved in obtaining useful results for more realistic problems and algorithms often leads to an empirical evaluation of the candidate algorithms. The experimental analysis of optimization metaheuristics should (1) provide users with recommendations about which technique to apply in a particular situation, as well as (2) provide designers with insight on the workings and effectiveness of different algorithmic components (and their combinations) in different classes or instances of problems.

No literature survey will be attempted here as there is a rich literature on designing, evaluating, and reporting on computational experiments in general [11] and with heuristics in particular [10, 3, 13, 5, 6], as well as building test-suites [15] and performing comparative studies [14, 9].

Denoting the cardinality of a set A by $|A|$, the most common way of assessing the relative performance of a set S of solvers $s_j, j \in \{1, \ldots, |S|\}$ is to define a set P of "representative" problems $p_i, i \in \{1, \ldots, |P|\}$ and then test all solvers against all problems measuring the performance $t(p_i, s_j)$ of solver $s_j \in S$ when applied to problem $p_i \in P$. Two important tasks arise now: (1) the choice of the set of problems P, and (2) the definition of a meaningful performance measure.

It is not an easy task to define a set of problems which is representative of the domain of interest, as one would like it to (1) span the target problem-space and, at the same time, (2) be as small as possible, in order to alleviate the computational burden associated with the experiments. It is also clear that, as the number of test-problems grows, the amount of data produced may reach the point of making its visualization, interpretation and analysis a hard task.

The first step in order to evaluate $t(p_i, s_j)$ is to define a meaningful goal and then measure the amount of resources required by the solver to achieve that goal. Examples of goals include (1) to reach a given objective function level $\bar{f}$: that means finding an $\bar{x}$ such that $f(\bar{x}) \leq \bar{f}$ $\left(f(\bar{x}) \geq \bar{f}\right)$ for a minimization (maximization) problem, and (2) to find a feasible solution: for a constrained optimization problem, that means finding $\bar{x}$ such that $g_q(\bar{x}) \geq 0$, $q = 1, 2, \ldots, \bar{q}$ and $h_q(\bar{x}) = 0$, $q = \bar{q}+1, \ldots, m$. Examples of resources include (1) CPU time and (2) number of objective function evaluations (useful for black-box optimization involving expensive simulations).

Although many performance measures have been suggested in the literature [5], in this paper we are particularly interested in performance profiles [8], which are analytical tools to facilitate the visualization and interpretation of the results of benchmark experiments. Some useful alternative performance profiles are also presented here as well as alternative ways to compare the solvers.

Although the benchmarking process is usually mainly intended to provide an assessment of the relative performance of the solvers/algorithms considered in the experiment, it would also be interesting to assess the relative performance of the test-problems themselves with respect to the solvers. Are all test-problems relevant to the final result? Would it be possible to remove some of them without affecting the relative performance of the solvers (no changes in their final ranking)? Are some test-problems too easy (or too difficult) so that they do not have the ability to help to discriminate the solvers? We believe that these questions can, at least partially, be answered by taking another look at the same data used to build the solvers's performance profiles.

In the next section, performance profiles are presented. In Sect. 3 we discuss some performance profile variants. Section 4 shows how performance profiles can be defined in order to assess the discriminative power of each test-problem of the suite. We illustrate the application of the proposed extensions of performance profiles using real data from a competition comparing nature-inspired metaheuristics. The paper ends with some conclusions and suggestions for future work.

2 Performance Profiles

In order to introduce performance profiles, consider that $t(p_i,s_j)$ is the CPU time spent by solver s_j to reach a pre-defined objective function value in problem p_i. If a solver fails to reach the goal for a given problem, the corresponding $t(p_i,s_j)$ is set to a conveniently defined large number. Given the definition of $t(p_i,s_j)$, we introduce the performance ratio as

$$r(p_i,s_j) = \frac{t(p_i,s_j)}{\min\{t(p_i,s_k) : s_k \in S\}}. \tag{1}$$

Although each $t(p_i,s_j)$ or $r(p_i,s_j)$ is worth considering by itself, it would be convenient to be able to assess the performance of the solvers in S on a large set of problems P in a graphical form suitable for human inspection. This has been done by Dolan and Moré, who introduced performance profiles [8]. Defining

$$\rho_{s_j}(\tau) = \frac{1}{|P|}\left|\{p_i \in P : r(p_i,s_j) \leq \tau\}\right| \tag{2}$$

then $\rho_{s_j}(\tau)$ is the fraction of problems in P that $s_j \in S$ is able to solve within a factor $\tau \geq 1$ of the best observed performance. If the set P is representative of the class of problems of interest then solvers with larger $\rho_{s_j}(\tau)$ are to be preferred.

From the definition above it follows that the performance profile $\rho_{s_j} \in [0,1]$ for solver s_j is a nondecreasing, piecewise constant function, continuous from the right at each discontinuity point. It has also be shown [8] that it is insensitive (1) to the results on a small set of problems, and (2) to small variations in several problems. In addition we also point out that the area under the ρ_{s_j} curve

$$AUC_{s_j} = \int \rho_{s_j}(t)\,dt$$

is an overall performance measure for solver s_j in the problem set P [1].

Another desirable feature of a solver is its *reliability*, which can also be assessed from the performance profiles. A measure of the reliability of solver s is its relative performance in the problem where it performed worst [1]

$$R_s = \sup\{\tau : \rho_s(\tau) < 1\}.$$

As a result, the most reliable solver is the one that minimizes R_s; that is, it presents the best performance, in the worst case, in the set P. This is in fact a minsup criterion:

$$s^* = \arg\min_{s \in S} R_s = \arg\min_{s \in S} \sup\{\tau : \rho_s(\tau) < 1\}$$

To illustrate the ideas above, we provide a simple example in Fig. 1 which shows hypothetical performance ratios and the corresponding performance profiles for solvers A, B, and C.

	A	B	C
P0	1.5	1.0	4.5
P1	2.0	1.0	4.2
P2	2.0	1.0	1.5
P3	2.2	1.0	1.1
P4	2.4	1.6	1.0
P5	2.5	1.8	1.0
P6	2.6	2.2	1.0
P7	2.8	2.4	1.0
P8	2.9	2.5	1.0
P9	3.0	4.0	1.0

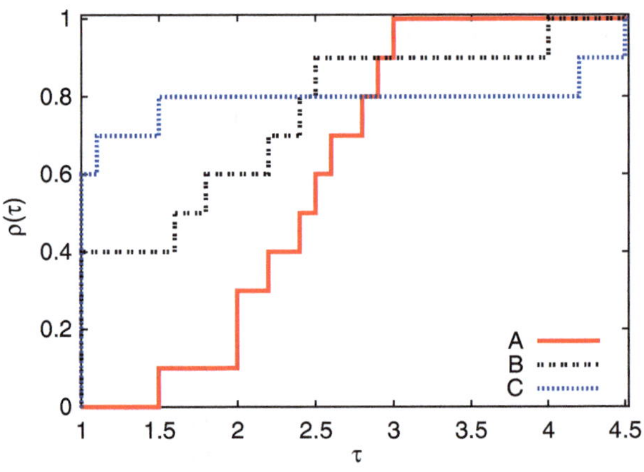

Fig. 1: An example of performance ratios and the performance profiles for solvers A, B, and C

In Fig. 1 it is easy to see that $\rho_C(1) > \rho_B(1) > \rho_A(1)$ and thus solver C is the one that comes out first more often in the set P. As $\rho_A(1) = 0$ solver A is never the best for the problems in P. Also, one has $AUC_C > AUC_B > AUC_A$, and thus solver C is the most efficient one, followed by B and A. Finally, solver A is the most reliable one, since it solves all problems in P within a factor of three from the best performance.

It should be noticed that probabilistic performance profiles for the experimental evaluation of stochastic algorithms have already been proposed [4].

3 Performance Profiles Variants

An underlying hypothesis present in [8] is that all problems in P have the same importance and, consequently, they have the same influence in the relative performance assessment. That may not always be the case as it could be previously agreed that some problems should have a higher influence in the performance evaluation [2].

As a result, we suggest here the introduction of a unique weight for each test-problem as a means of more accurately assessing the relative performance of the solvers considered. The first proposal assumes that solving a larger problem (one with a larger number of variables) should be more rewarding than solving a smaller one. As a result, and denoting the dimension of problem p_i by n_{p_i}, one could use the *relative number of degrees of freedom*

$$w_{p_i} = \frac{n_{p_i}}{\sum_{k=1}^{|P|} n_{p_k}} \tag{3}$$

as the weight for the i-th test-problem. The performance profiles are now defined as

$$\rho_{s_j}(\tau) = \sum_{i=1}^{|P|} w_{p_i} I(r(p_i, s_j) \leq \tau) \tag{4}$$

where $I(r(p_i, s_j) \leq \tau)$ equals one if the condition holds, and zero otherwise. Instead of "counting problems solved" as in the original case, in which the weights are equal, one is now "counting variables solved for".

Another possibility is to define the weights *a posteriori*. After all the testing is done, the minimum amount of CPU time required by each problem is recorded. Then the weight for the i-th test-problem is given by

$$w_{p_i} = \frac{\min\{t(p_i, s_j) : s_j \in S\}}{\sum_{k=1}^{|P|} \min\{t(p_k, s_j) : s_j \in S\}} \tag{5}$$

which is the relative (minimum) CPU time to solve the i-th test-problem. Expression (5) will assign more importance to those problems that proved to be harder to solve. The performance profiles are again defined by Eq. (4).

Notice that the variants presented above change the vertical steps in the performance profile plots. This indicates that the weights have a global influence in the comparisons, when the solvers are to be ranked according to the area under their curves. For a given solver s_j, when the performance of the problem with smaller τ is vertically scaled, this variation affects the area not only under this part of the curve but also the area under the remaining ones.

3.1 A Multicriterion View of the Performance Comparison Problem

It would be interesting to find a way not only of indicating which solver in S has the best performance in the set of problems P, but also possibly ranking all solvers from best to worst. In fact one can look at the relative performance $r(p_i, s_j)$ of the j-th solver in the i-th test-problem as an independent objective, or criterion, to be optimized. Each solver s_j can be seen as a point of coordinates $r(p_i, s_j) = r_i(s_j), i = 1, \ldots, |P|$ in the relative performance space $R^{|P|}$. The ideal best solver ($s_j = s_{j^*}$) would be the best performer in every test-problem in P, that is, $r(p_i, s_{j^*}) = 1$ for all i. However, this is very hard to be accomplished in a large set of representative problems. If solver s_1 is the best in a subset of problems from P while s_2 is the best in the remaining problems, how to decide that one is better than the other?

To exemplify, and to allow for a simple graphical representation, one could think of a hypothetical benchmark situation with only two test-problems ($|P| = 2$) and four solvers, as sketched in Fig. 2.

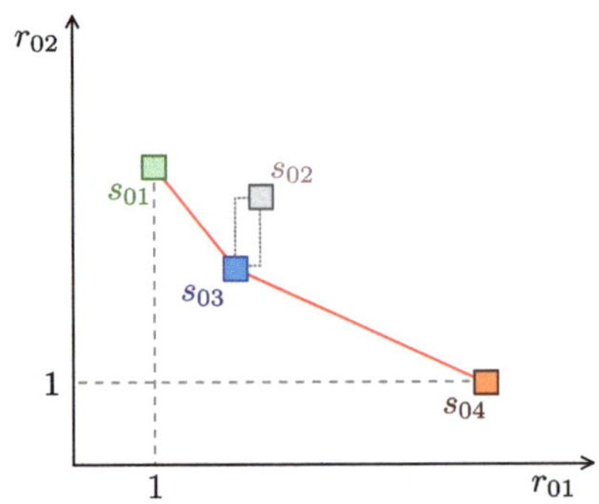

Fig. 2: Illustration of a multicriterion view considering the relative performance $r(p_i, s_j)$ on test-problems p_{01} and p_{02} as the two objective functions (to be minimized), and the relative performance of the solvers s_{01}, s_{02}, s_{03}, and s_{04} as the coordinates in this (relative) performance space $R^{|P|}$. Solver s_{01} is the best performer on problem p_{01}, and solver s_{04} is the best performer on problem p_{02}

In the parlance of multiobjective optimization [7] one can see that solver s_{02} is *dominated* by solver s_{03}, as solver s_{03} has a better performance in all ($|P| = 2$) test-problems considered. On the other hand, nothing can be said about dominance among solvers s_{01}, s_{03}, and s_{04}; they are non-dominated, belonging to the so-called Pareto front. It is clear then that the problem of choosing the best performer and ranking the remaining ones can be seen as a multicriterion decision-making process.

Thus, one could then take the same data used to construct the performance profiles and define a measure of quality of the solver s_j (the smaller the better, since $r(p_i, s_j) \geq 1$) as

$$V(s_j) = \prod_{i}^{|P|} r(p_i, s_j). \tag{6}$$

This is the volume of the hyperbox defined by s_j and the origin in $R^{|P|}$.

Notice that the definition above does not consider any differences between the problems, that is, similarly to the standard performance profiles, every problem has the same importance and is equally weighted. However, having a higher relative performance in a harder test-problem seems to be more relevant than having such good performance in an relatively easier test-problem. This can be taken care of by weighing the contribution of each problem with respect to its hardness by including the (normalized) hardness measure of the problem p_i as an exponent for each term $r(p_i, s_j)$ leading to

$$V(s_j) = \prod_{i}^{|P|} [r(p_i, s_j)]^{w_{p_i}} \tag{7}$$

with w_{p_i} from Eq. (5).

3.2 Illustrating with a Real Case

In order to illustrate the use of the proposed extensions of performance profiles, we have considered the results of the competition organized within the Special Session on Constrained Real-Parameter Optimization [14] held during the IEEE Congress on Evolutionary Computation in 2006, to evaluate nature-inspired metaheuristics.

A set of 24 minimization problems in R^n with equality as well as with inequality constraints was used. The dimension n_{p_i} of each one of the 24 test-problems p_i, which compose the benchmark of the CEC 2006 competition, is given in Table 1.

Table 1: Dimension of each CEC 2006 competition test-problem

Problem	p_{01}	p_{02}	p_{03}	p_{04}	p_{05}	p_{06}	p_{07}	p_{08}	p_{09}	p_{10}	p_{11}	p_{12}
Dimension	13	20	10	5	4	2	10	2	7	8	2	3
Problem	p_{13}	p_{14}	p_{15}	p_{16}	p_{17}	p_{18}	p_{19}	p_{20}	p_{21}	p_{22}	p_{23}	p_{24}
Dimension	5	10	3	5	6	9	15	24	7	22	9	2

The result of the competition was defined based on the computation, for each problem, of the following quantities: feasible rate, success rate, and success performance. For our purposes here it is sufficient to consider only the last one.

The names of the algorithms are omitted as we are only interested in applying the performance profiles to the corresponding (real case) data. As no technique was able to solve problems p_{20} and p_{22} in the CEC 2006 competition, those test-

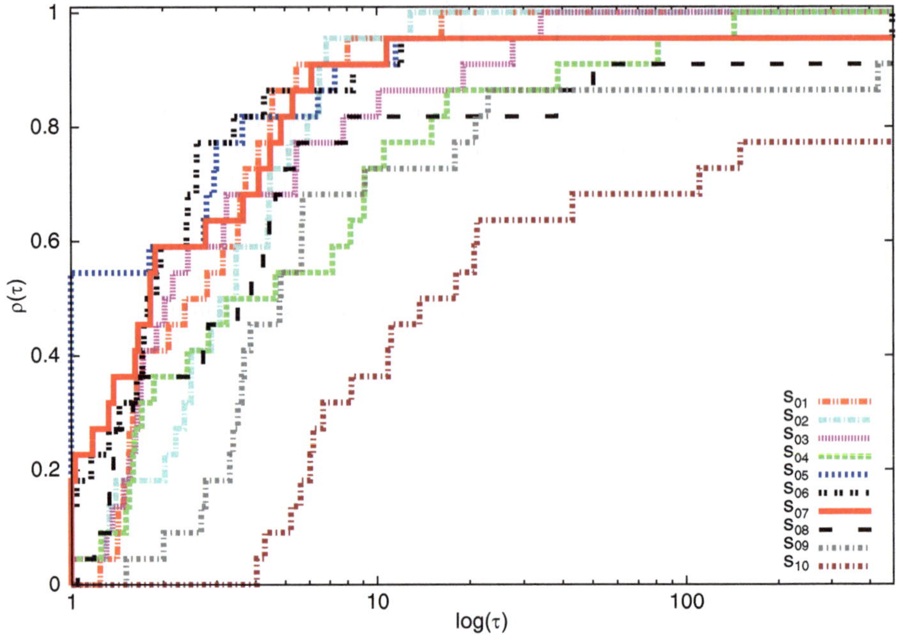

Fig. 3: Performance profiles for the results of the CEC 2006 competition (problems p_{20} and p_{22} were not included)

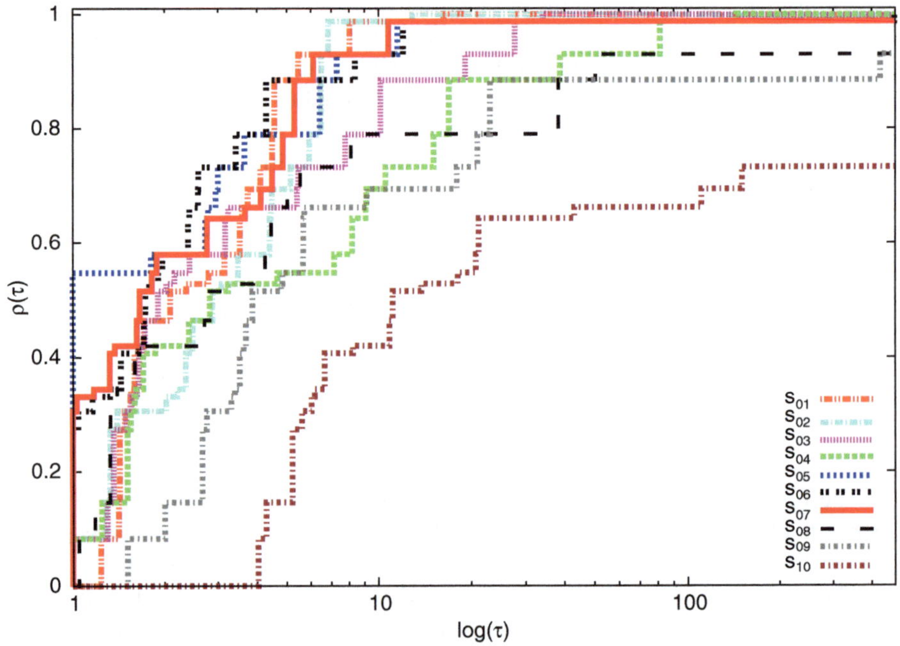

Fig. 4: Performance profiles for the results of the CEC'06 competition weighted by the dimension of the problems

problems can be excluded *a priori* from the analysis. The corresponding standard performance profiles are presented in Fig. 3. The performance profiles when the test-problems are weighted according to problem dimensionality (Eq. 3) and hardness (Eq. 5) are shown in Figs. 4 and 5, respectively. Table 2 shows the rank induced by the area under the curve considering the standard performance profile as well as its two weighted variants.

Table 2: Comparison among the rankings obtained using the different performance profile definitions

	1	2	3	4	5	6	7	8	9	10
Standard	s_{01}	s_{02}	s_{03}	s_{04}	s_{05}	s_{06}	s_{07}	s_{08}	s_{09}	s_{10}
Weighted by dimension	s_{06}	s_{05}	s_{01}	s_{07}	s_{02}	s_{03}	s_{04}	s_{08}	s_{09}	s_{10}
Weighted by hardness	s_{01}	s_{02}	s_{07}	s_{05}	s_{06}	s_{03}	s_{04}	s_{08}	s_{09}	s_{10}

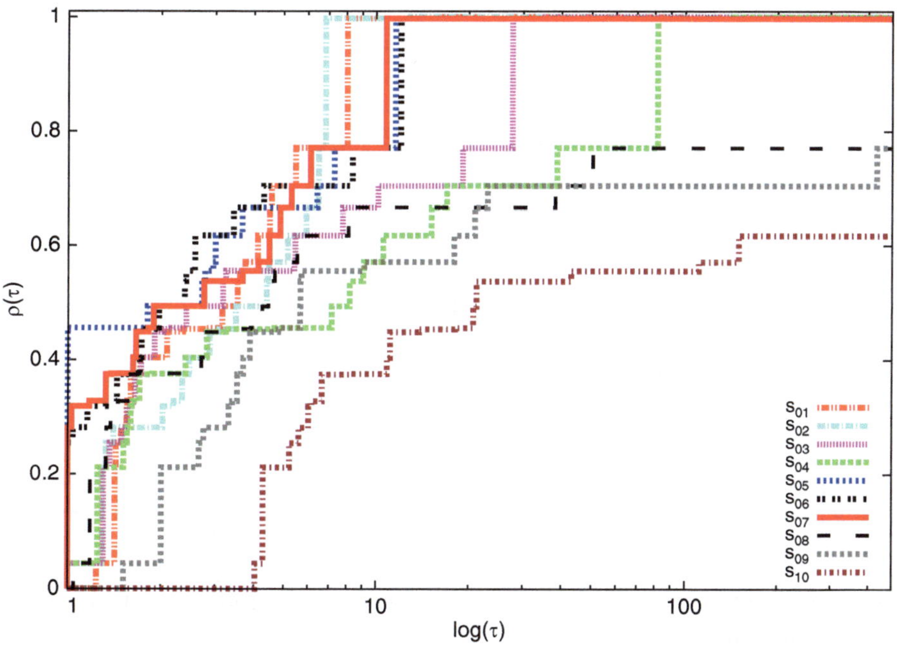

Fig. 5: Performance profiles for the results of the CEC'06 competition weighted by problem hardness

It can be seen that the different performance profiles definitions lead to different rankings of the solvers, except for the agreement on the three least-performing ones. Although the weighted profiles agree on the five least-performing solvers, the use of problem dimensionality as a weight leads to a quite different ranking of the top performers (see the change in s_{06}), indicating that for this set of problems dimensionality does not directly correlate with hardness. The proposed performance profiles weighted by problem hardness seem to be the most meaningful one, in spite of agreeing with the standard one—in this particular benchmark—with respect to the two top performers.

Finally, Table 3 presents the ranking of the solvers according to their hypervolumes in the relative performance space $R^{|P|}$. The ranking with respect to (1) the area under the curves when using the standard performance profiles, (2) the hypervolume of the results, and (3) the weighted hypervolumes are shown. One can verify that both rankings generated by the hypervolume are the same. Also notice that the preferable solvers when the hypervolume is considered are, more often than not, the same ones when the area under the curve of the performance profiles (weighted by dimensionality of the problem) are considered.

Table 3: Comparison among the rankings obtained using the performance profiles and the hypervolume

	1	2	3	4	5	6	7	8	9	10
Standard	s_{01}	s_{02}	s_{03}	s_{04}	s_{05}	s_{06}	s_{07}	s_{08}	s_{09}	s_{10}
Hypervolume	s_{05}	s_{06}	s_{01}	s_{07}	s_{02}	s_{03}	s_{08}	s_{04}	s_{09}	s_{10}
Weighted hypervolume	s_{05}	s_{06}	s_{01}	s_{07}	s_{02}	s_{03}	s_{08}	s_{04}	s_{09}	s_{10}

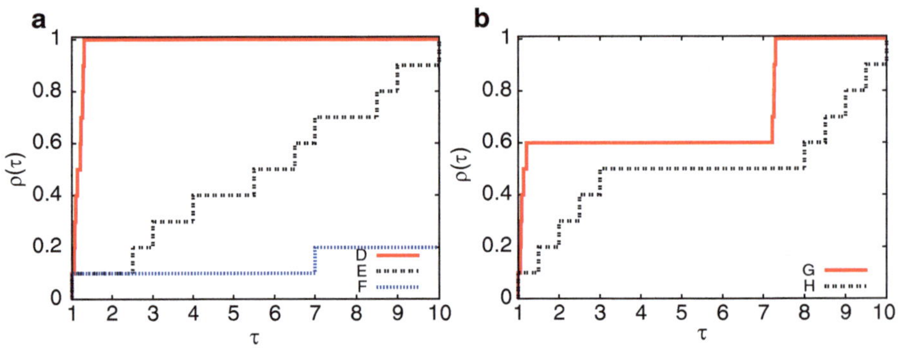

Fig. 6: Examples of performance profiles for hypothetical problems D, E, and F (**a**); and G and H (**b**)

4 Evaluating the Suite with Performance Profiles

We argue that it is possible to examine the same data that is collected to construct performance profiles and then, for each test-problem, to plot a curve that provides useful information regarding the behavior of this particular test-problem with respect to the set of solvers considered. Considering the table containing the relative performance indexes $r(p_i, s_j)$ (Eq. 1) used to build the performance profiles (Eq. 2), one can define

$$\gamma_{p_i}(\tau) = \frac{1}{|S|} \left| \{s_j \in S : r(p_i, s_j) \leq \tau\} \right| \tag{8}$$

which can be thought of as the analogue, for the i-th test problem, of the standard performance profile of a given solver. As a result, $|P|$ curves can be plotted and compared. An illustrative example, Fig. 6, displays the graphs for $\gamma_{p_i}(\tau)$ corresponding to hypothetical test-problems (a) D, E, and F, and (b) G and H applied to ten solvers.

It is easy to see in Fig. 6a that test-problem D does not clearly discriminate among the solvers in S; they are all able to solve this test-problem with very similar performances. Test-problem F, on the other hand, seems too difficult, as only two algorithms in S are able to solve it. No information is provided regarding the performance of the remaining solvers. Test-problem E provides better discrimination, as different solvers achieve different levels of performance. Finally, in Fig. 6b, the test-problems G and H are able to cluster the solvers in S in two groups, one of them achieving a higher performance on this test-problem than the other. Also, it can be seen that test-problem H has a higher discriminative power within each cluster of solvers.

Besides visual inspection, one can also define a way to automatically identify the test-problem(s) with low discriminative power. For a given solver, one can compute the sum of the differences between the values of τ associated with each test-problem and the (ideal) even distribution along the τ axis. Thus, test-problems with plots more similar to the uniform distribution are preferable while those with high discrepancy are candidates to be removed. However, it is important to note that features similar to those observed in Fig. 6b are only detected after the runs are performed. In any case, it is felt that this is an easy and practical way of removing problems with low discriminative power, and possibly substituting them, thus assisting the design of better sets of test-problems. In the following section, this proposal is applied to the results of a real case.

4.1 Illustrating with a Real Case

Using the same data of Sect. 3.2, the performance profiles for the test-problems with respect to the solvers in the CEC 2006 competition are presented in Fig. 7. The curves are presented separately to improve readability. A ranking of the problems according to the minimization of the difference between its values of τ and the values uniformly distributed in the interval is shown in Table 4.

Now one can think of answering the questions posed earlier in the Introduction. It is easy to see that test-problems p_{20} and p_{22} are too difficult (no solver is able to solve them) and thus they do not affect the rank of the algorithms. Also, by the plots

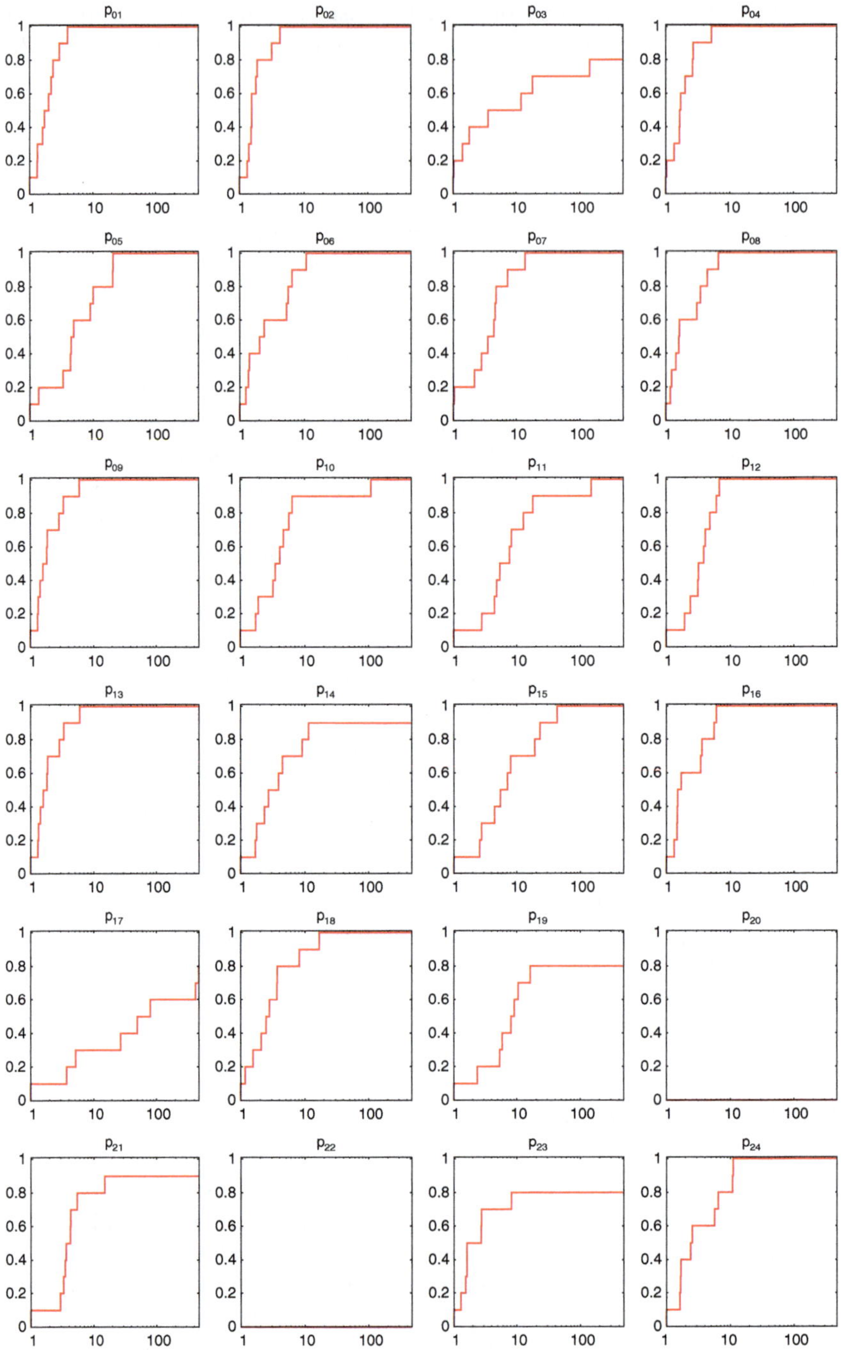

Fig. 7: Performance profiles for the problems of the CEC'06 competition

Table 4: Ranking of the problems

Rank	1	2	3	4	5	6	7	8	9	10	11
Problem	p_{17}	p_{03}	p_{19}	p_{23}	p_{13}	p_{21}	p_{14}	p_{11}	p_{10}	p_{15}	p_{05}
Rank	12	13	14	15	16	17	18	19	20	21	22
Problem	p_{07}	p_{24}	p_{18}	p_{06}	p_{12}	p_{16}	p_{08}	p_{09}	p_{04}	p_{01}	p_{02}

in Fig. 7, one can notice that test-problems p_{01}, p_{02}, p_{04}, and p_{09} do not provide a good discrimination among the solvers in S. Those observations suggest that perhaps it would be possible to remove those test-problems from the suite P. In fact, by doing so one obtains the performance profiles in Fig. 8. By computing the final ranking (considering the areas under the curves) no changes in the ranking are observed. On the other hand, by removing a single test-problem with high discriminating power such as p_{17}, one obtains the performance profile shown in Fig. 9, and a very different ranking of the solvers.

Table 5 presents (1) the original ranking, (2) the ranking obtained by removing p_{01}, p_{02}, p_{04}, and p_{09}, and (3) the ranking resulting from removing only p_{17}. Recall that, in all cases, test-problems p_{20} and p_{22} have been excluded from the suite.

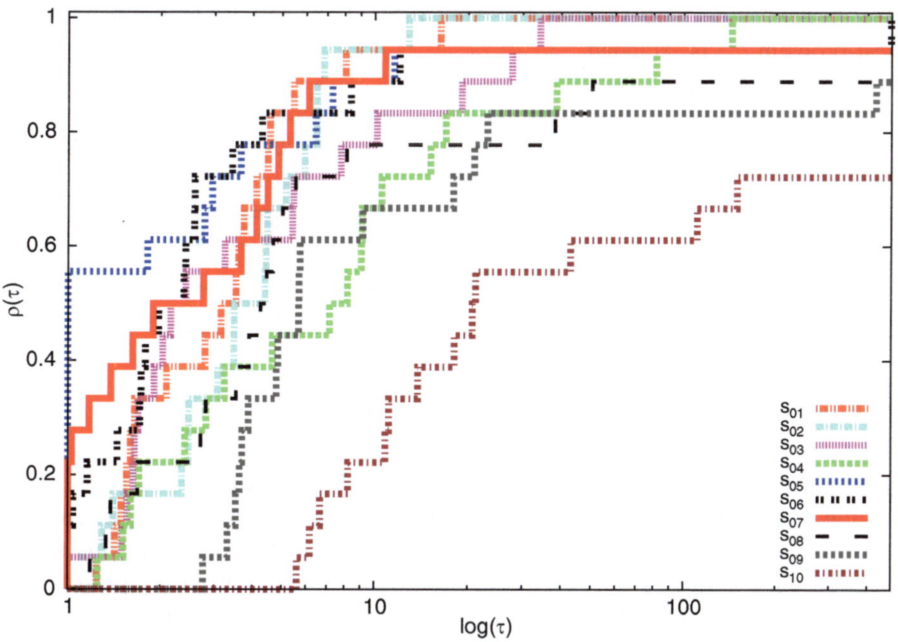

Fig. 8: Performance profiles for the results of the CEC 2006 competition when the problems p_{01}, p_{02}, p_{04}, and p_{09} are not considered

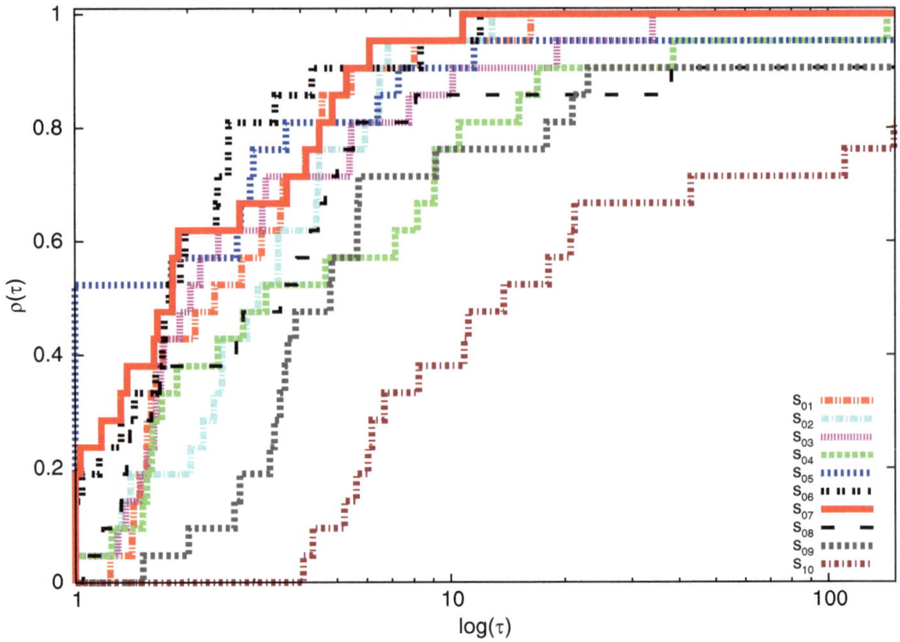

Fig. 9: Performance profiles for the results of the CEC 2006 competition when the test-problem p_{17} is excluded

Table 5: Comparison between the original ranking (with respect to the AUC) and the ranking obtained when different test-problems are removed from the suite

	1	2	3	4	5	6	7	8	9	10
Standard	s_{01}	s_{02}	s_{03}	s_{04}	s_{05}	s_{06}	s_{07}	s_{08}	s_{09}	s_{10}
AUC from Fig. 8	s_{01}	s_{02}	s_{03}	s_{04}	s_{05}	s_{06}	s_{07}	s_{08}	s_{09}	s_{10}
AUC from Fig. 9	s_{06}	s_{07}	s_{01}	s_{02}	s_{03}	s_{05}	s_{04}	s_{08}	s_{09}	s_{10}

5 Conclusions

Benchmark experiments play a fundamental role in the design, analysis, and evaluation of algorithms in general and of optimization metaheuristics in particular. In this paper we discussed performance profiles, which are tools for analyzing the results of such experiments. We argued that, due to their explanatory power, such profiles should be more widely used by the metaheuristics community.

We have shown two performance profiles variants, weighted by (1) problem dimensionality and (2) problem hardness. These variants can be extracted from the data collected during the experiment and provide additional views on the same experiment. Also, a multicriterion view of the performance comparison problem was presented illustrating the evaluation of a set of solvers from a different angle.

Besides that, we have shown that test-problem performance profiles can be built from the same performance profiles data, displaying the relative relevance (discriminatory power) of each test-problem in the context of the benchmarking suite and with respect to the set of solvers considered. Also, a way to automatically detect the problems which are candidates to be removed from the benchmarking suite was proposed.

In order to illustrate their potential, we used performance profiles to analyze the results of previous work in the literature, which suggested that (1) the performance profiles weighted by hardness are preferable; (2) the multicriterion point of view is interesting for the performance comparison problem; (3) the inclusion of the hardness in the decision based on multicriterion did not affect the ranking; and (4) the performance profiles proposed here for test-problems were able to indicate the most, as well as the least, relevant problems in a suite.

A possible direction for future work is to apply the ideas presented in this paper to other variants of the performance profiles, such as the data profiles presented by Moré and Wild [12], in order to better assess the behavior of both solvers and test-problems involved in benchmarking experiments.

Acknowledgements The authors thank the reviewers for their comments, and LNCC, CNPq (grant 311651/2006-2), and FAPERJ (grants E-26/ 102.825/2008 and E-26/100.308/2010) for their support.

References

1. Barbosa, H.J.C., Bernardino, H.S., Barreto, A.M.S.: Using performance profiles to analyze the results of the 2006 CEC constrained optimization competition. In: Proceedings of the World Congress on Computational Intelligence – CEC, pp. 1–8. IEEE, Barcelona, Spain (2010)
2. Barbosa, H.J.C., Bernardino, H.S., Barreto, A.M.S.: Exploring performance proles for analyzing benchmark experiments. In: Proceedings of the Metaheuristics International Conference (MIC), Udine, Italy (2011)
3. Barr, R., Golden, B., Kelly, J., Resende, M., Stewart, W.: Designing and reporting on computational experiments with heuristic methods. J. Heuristics **1**, 9–32 (1995)
4. Barreto, A.M.S., Bernardino, H.S., Barbosa, H.J.C.: Probabilistic performance profiles for the experimental evaluation of stochastic algorithms. In: Proceedings of the Genetic and Evolutionary Computation Conference – GECCO, pp. 751–758. ACM (2010)
5. Bartz-Beielstein, T.: Experimental research in evolutionary computation – the new experimentalism. In: Natural Computing Series. Springer, Berlin (2006)
6. Bartz-Beielstein, T., Chiarandini, M., Paquete, L., Preuss, M. (eds.): Experimental Methods for the Analysis of Optimization Algorithms. Springer, Berlin (2010)
7. Deb, K.: Multi-Objective Optimization Using Evolutionary Algorithms. Wiley-Interscience Series in Systems and Optimization. Wiley, Chichester (2001)
8. Dolan, E., Moré, J.J.: Benchmarcking optimization software with performance profiles. Math. Program. **91**(2), 201–213 (2002)
9. Hansen, N., Auger, A., Ros, R., Finck, S., Pošík, P.: Comparing results of 31 algorithms from the black-box optimization benchmarking BBOB-2009. In: Proceedings of the Conference on Genetic and Evolutionary Computation, pp. 1689–1696, New York, NY, USA (2010)
10. Hooker, J.: Testing heuristics: We have it all wrong. J. Heuristics **1**, 33–42 (1995)

11. Jackson, R.H.F., Boggs, P.T., Nash, S.G., Powell, S.: Guidelines for reporting results of computational experiments. Report of the ad hoc committee. Math. Program. **49**, 413–425 (1991)
12. Moré, J.J., Wild, S.M.: Benchmarking derivative-free optimization algorithms. SIAM J. Optim. **20**(1), 172–191 (2009)
13. Rardin, R.L., Uzsoy, R.: Experimental evaluation of heuristic optimization algorithms: a tutorial. J. Heuristics **7**, 261–304 (2001)
14. Suganthan, P.N.: Special session on constrained real-parameter optimization (2006). http://www3.ntu.edu.sg/home/epnsugan/index_files/CEC-06/CEC06.htm
15. Whitley, L.D., Mathias, K.E., Rana, S.B., Dzubera, J.: Building better test functions. In: Proceedings of the 6th International Conference on Genetic Algorithms, pp. 239–247. Morgan Kaufmann, San Francisco, CA, USA (1995)

Real-World Parameter Tuning Using Factorial Design with Parameter Decomposition

Aldy Gunawan, Hoong Chuin Lau, and Elaine Wong

Abstract In this paper, we explore the idea of improving the efficiency of factorial design for parameter tuning of metaheuristics. In a standard full factorial design, the number of runs increases exponentially as the number of parameters. To reduce the parameter search space, one option is to first partition parameters into disjoint categories. While this may be done manually based on user guidance, an automated approach proposed in this paper is to apply a fractional factorial design to partition parameters based on their main effects where each partition is then tuned independently. With a careful choice of fractional design, our approach yields a linear rather than exponential run time performance with respect to the number of parameters. We empirically evaluate our approach for tuning a simulated annealing algorithm that solves an industry spares inventory optimization problem. We show that our proposed methodology leads to improvements in terms of the quality of solutions when compared to a pure application of an automated parameter tuning configurator ParamILS.

1 Introduction

The performance of a metaheuristic algorithm largely depends on the expertise in tuning the algorithm's control parameters. For example, a simulated annealing algorithm yields good solutions only if several parameters such as initial temperature, cooling factor, number of iterations and so on are properly tuned. However, finding the best combination of parameter settings is a tedious and time-consuming task.

Aldy Gunawan · Hoong Chuin Lau
School of Information Systems, Singapore Management University, Singapore
e-mail: aldygunawan@smu.edu.sg, hclau@smu.edu.sg

Elaine Wong
Innovation Works, EADS Singapore, Singapore
e-mail: elaine.wong.kl@eads.net

L. Di Gaspero et al. (eds.), *Advances in Metaheuristics*, Operations Research/ Computer Science Interfaces Series 53, DOI 10.1007/978-1-4614-6322-1_3,

In recent years, several automated approaches for finding good parameter settings have been proposed. These approaches can be classified into model-free algorithm configuration methods and model-based approaches. Model-free algorithms are simpler to implement than model-based approaches because the former can be applied out-of-the-box, while the latter requires iterations between fitting models and using them to make choices about which configurations to investigate [13]. Examples of model-free algorithm configuration methods are F-Race [4] and Iterated F-Race [5], ParamILS [10] and genetic algorithm GGA [2]; examples of model-based approaches are SPO+ [12] and SMAC [13]. SPO+ is an extension of the Sequential Parameter Optimization (SPO) framework, which focuses on tuning algorithms with continuous parameters for a single problem instance. SMAC, unlike SPO+, can be used to handle categorical parameters. Both, model-free and model based approaches have been used to optimize various algorithms for some classical combinatorial optimization problems, such as the propositional satisfiability problem (SAT), the travelling salesman problem (TSP) and MIP problems.

Many real-world problems make use of optimization algorithms that contain a set of parameters. For example, the CPLEX solver has 76 parameters that affect the search process [11]. In such problems, parameter search space reduction can be important in order for automated tuning to become computationally feasible. By decomposing the parameters into disjoint partitions and tuning them separately, the parameter search space will be significantly reduced. Lau and Xiao [14], for example, decompose the parameters set into disjoint graphs based on the correlation among the parameters, and the approach was used for tuning a genetic algorithm (GA) for the bandwidth minimization problem (BMP).

Another approach for decomposing parameters is found in Design of Experiments (DOE). DOE has been used to systematically find the best parameters values for a heuristic. Coy et al. [7] proposed a procedure, based on DOE and gradient descent, to find parameter settings for vehicle routing heuristics. The drawbacks of the approach are that: the linear approximation of the response surface and the average setting might not be appropriate if the class of problems is too broad. Adenso-Díaz and Laguna [1] developed CALIBRA, which employs a Taguchi fractional experimental design and a local search procedure to tune up to five parameters. CALIBRA only focuses on the linear assumption without examining interactions between parameters. Hutter et al. [13] proposed the use of response surface models to characterize the importance of parameters for future research direction. All the above-mentioned approaches show that factorial experimental design is particularly useful in the early stages of experimental work, when many parameters are likely to be investigated. Unfortunately, the caveat of a full factorial design is that the number of runs required increases exponentially with the number of parameters [8, 16].

In this paper, we propose a decomposition approach to reduce the parameter space of parameters. Consider an algorithm, called the target algorithm, that requires a number of parameters to be tuned. We divide the parameters into a number of disjoint categories. This is done either manually based on user guidance; or automatically using a fractional factorial design, which measures the main effect ranks among the parameters. Since our focus is to separate main effects and interactions, the so-called Resolution IV Design [15] is used.

We then apply the proposed tuning framework by Gunawan, Lau, and Lindawati [8] for each category of parameters. The proposed tuning framework is divided into three phases, namely screening, exploration and exploitation. Given a set of parameters to be tuned, the screening phase seeks to rank these parameters so as to determine unimportant parameters, whose values have insignificant impact on the solution quality. Values for unimportant parameters can be set to some constant numbers and thereby further reduce the resulting parameter space to be explored. In the exploration phase, a first-order polynomial model based on a response surface is then built to define the promising initial ranges for the important parameters. These promising initial ranges are then sent to an automated tuning configurator such as ParamILS [10] in the exploitation phase to find the desired parameter setting.

This work arises from a real optimization problem in the aerospace company EADS. We are concerned with tuning an existing algorithm that involves a set of parameters and intensive computations. The major contributions of this paper are summarized as follows:

- We propose the idea of parameter decomposition and describe two approaches.
- We apply our approach to tune a simulated annealing algorithm that is used to solve a computationally-intensive spares inventory optimization problem.

The remainder of this paper is organized as follows. The automated tuning framework of [8] is outlined in Sect. 2. Section 3 presents the spares inventory optimization problem and the target algorithm. Section 4 presents our decomposition approaches. Section 5 provides computational results of our proposed approach applied to the spares inventory optimization problem. Finally, we provide some concluding perspectives and future research directions in Sect. 6.

2 Automated Tuning Framework

The automated tuning problem is defined as follows: θ is the finite set of candidate parameter configurations; X is a set of parameters to be tuned (each parameter can be either discrete or continuous, over a numeric range); TA is a target algorithm; I_{tr} is a set of training instances; $H(x)$ is a function that measures algorithm $TA's$ performance under a fixed parameter setting x on a set of problem instances. The solution of the automated tuning problem is the configuration x^* such that:

$$x^* = argmin_x H(x) \tag{1}$$

In this paper, we define the performance metric as the average percentage deviation of the set of obtained solutions from the optimal (or best known) solutions. The goal of automated tuning is to find x that optimizes the performance metric w.r.t. an (unknown) instance distribution. While the true performance metric cannot be computed exactly, it can be estimated using the given set of training instances I_{tr}. Subsequently, to verify the quality of this parameter setting, we measure the performance against a set of test instances.

The framework proposed in [8] consists of three phases, namely screening, exploration and exploitation. In the screening phase, we determine which parameters exhibit significant main effects thereby reducing the number of parameters to consider. Parameters with statistically significant p-value (less than 10 %) are referred to important parameters, while parameters with p-value greater than 10 % are considered unimportant parameters. For this purpose, [8] proposes a 2^k factorial design which consists of k parameters, where each parameter p_i only has two levels (l_i and u_i). A complete design requires $(2 \times 2 \times \cdots \times 2) \times a = a \times 2^k$ observations where a represents the number of replicates for a particular parameter setting (see [15]).

Let m be the total number of important parameters ($m \leq k$) determined in the screening phase. The exploration phase runs the target algorithm with respect to $2^m + 1$ possible parameter settings with an additional parameter setting defined by the centre point value of each parameter. A first-order (planar) model is then built to represent the relationship among parameters and the objective function value.

In order to test the significance of the planar model, interaction and curvature tests have to be conducted. The interaction test is used to test the significance of any interaction between parameters by looking at the significance of the estimated coefficient between two parameters. The curvature test tests whether the planar model is adequate to represent the local response function. The surface of parameters can still be approximated by a planar model as long as the existence of either interaction or curvature is not significant.

If the planar model assumption still holds, the process is then continued by applying steepest descent that moves rapidly to the vicinity of the optimum. From a statistical point of view, the region close to the optimum has been reached if the planar model assumption does not hold anymore. The final range for each important parameter is used as the input in the exploitation phase in which a configurator is used to find the optimal point in the region. In this study (as in [8]), ParamILS is applied to tune the target algorithm.

3 Case Description

In this section, we provide a description of the target algorithm used in this paper. Our problem is an aircraft spares inventory optimization problem arising from maintenance, repair and overhaul (MRO) operations faced by the aircraft total service support provider. The service provider is required to provide necessary spares to meet target service levels of customers of performance-based contracts operating out of a network of airports. The goal is to determine the optimal inventory allocation strategy that can fulfill target services levels. Optimality is defined in terms of total life cycle costs for spares comprising inventory holding cost, part purchasing and repair cost, logistics delivery cost, while service levels in terms of spares fill-rates.

The inclusion of logistic policies and strategies into the inventory optimization model allow better representation of real world operations. This problem incorporates (a) the use of lateral transshipment between warehouses, (b) service-time commitments for spares delivery, (c) inventory rationing for selected customers, and (d) option to scrap non-serviceable parts. Lateral transshipments are used when a serviceable spare part is not available at a local warehouse where the demand is triggered. In such cases, the serviceable spare part is shipped from an alternative warehouse to fulfill the urgent demand. As a result, the one (serviceable)-for-one (unserviceable) exchange would incur logistics transport costs. The optimized inventory allocation is required to make strategic trade-offs between dispatching from a central warehouse (lower warehousing costs) versus distributing inventory across the entire network (lower logistics transport costs) (For further reference, refer to [3]).

Service-time commitments are applicable in cases where spares demand is met from an alternative warehouse via lateral transshipments. The logistics transport has to deliver the serviceable part in the requested time in order to fulfill the demand. Failing to do so results in a penalty imposed on the service provider. Higher penalty rates encourage tighter conformance to the target service level, and possibly to the extent of even surpassing the targets required by the customers. Inventory rationing is a strategy adopted to prioritize different classes of customers. Unlike a simple pooling strategy, this strategy involves reserving inventory for customers requiring higher target service levels, as part of a higher priced contract and higher penalty rates [6]. Scrapping of parts is a resupply strategy to handle unserviceable parts. When deciding on scrap rates, the service provider takes into account repair shop turn-around times and costs, as well as a new part unit price and purchase lead time. The above problem is solved by a Simulated Annealing algorithm (see Fig. 1).

The parameters that need to be tuned are summarized in Table 1. The SA solution-acceptance criteria, which we term as oracle function, accepts a solution (x_k) that is worse than the incumbent solution (x) with the probability $Pr(x, x_k, T_k) = e^{(F(x)-F(x_k))/T_k}$, where T_k is the temperature at iteration k. In the oracle function, the exponential function is close to one if the new objective value $F(x_k)$ is close to $F(x)$, and approaches zero as the difference increases. An additional parameter oracle strictness ($Oracle$) is introduced into the SA algorithm. Oracle strictness adjusts the probability of acceptance as follows:

$$\hat{Pr}(x, x_k, T_k) = \left\{ \frac{Pr(x, x_k, T_k) - Oracle/100}{1 - Oracle/100} \right\}^{+} \tag{2}$$

The probability of accepting solution x_k is zero if $Pr(x, x_k, T_k)$ is less than $Oracle$, hence it is possible to tune the algorithm to a simple local search algorithm by setting $Oracle$ to 100. Unlike the cooling function, which dynamically adjusts T_k and hence the likelihood of acceptance according to the iteration count, the oracle strictness is fixed throughout the algorithm.

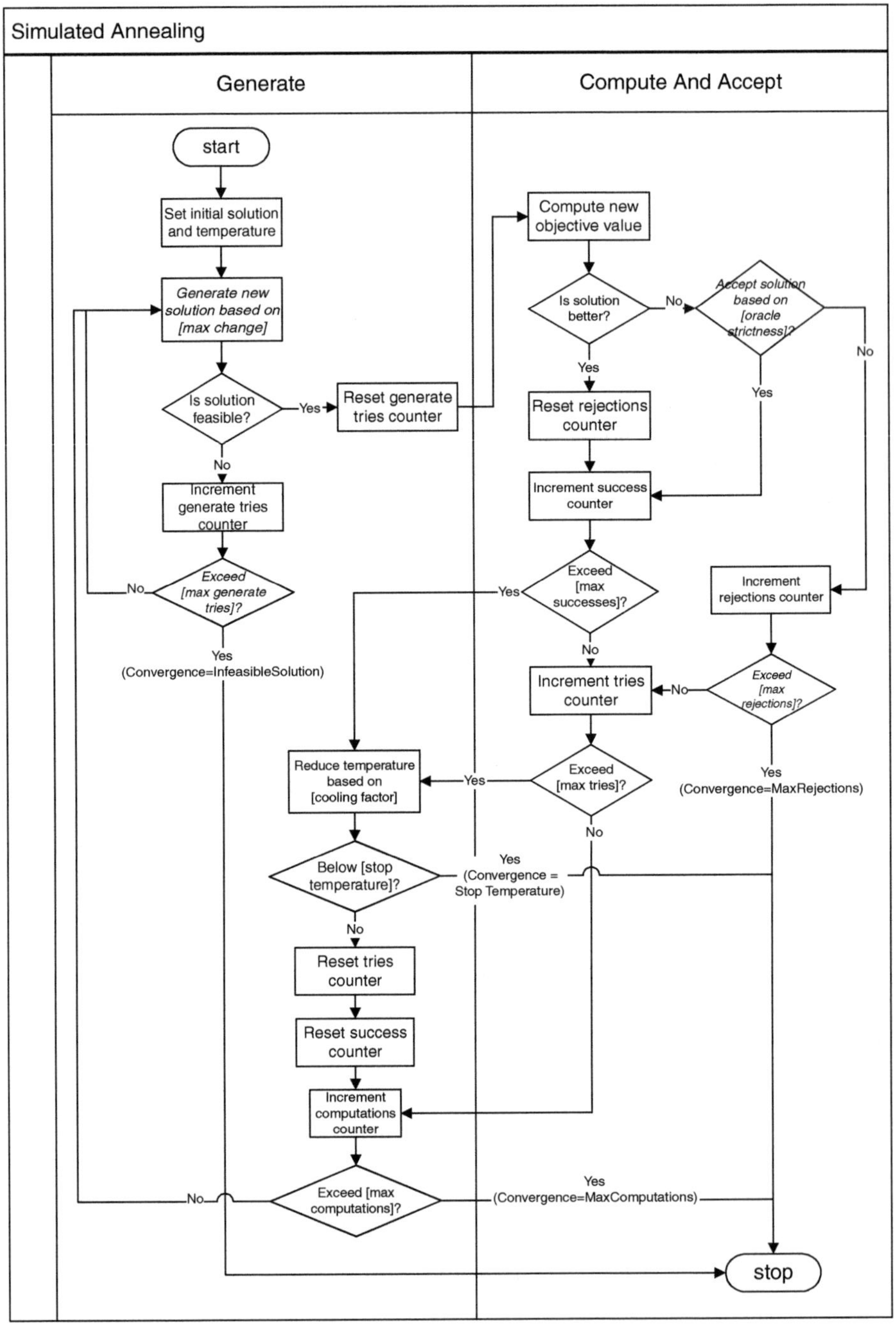

Fig. 1: Flow chart of simulated annealing algorithm

Table 1: Parameter definition

Parameters (p_i)	Abbreviation	Definition
Max successes	Success	Maximum number of successes within one temperature
Max tries	Tries	Maximum number of tries within one temperature
Max computations	Comp	Maximum number of solutions generated
Max rejections	Reject	Maximum number of consecutive rejections
Max change	Change	Maximum change in a variable value when generating a new solution
Max generate tries	Generate	Number of tries to generate a feasible solution
Cooling factor	Cooling	Factor to reduce the temperature by during each Temperature change
Oracle strictness	Oracle	A value to depict the strictness of the oracle function in accepting a new solution that has an objective value worse than the current one. A higher value would result in a higher rejection rate (e.g. a value of 100 would accept only better solutions)

4 s-step Decomposition Approach

The main idea of the s-step decomposition approach is to decompose n parameters into s disjoint categories, namely CAT_1, CAT_2, ... CAT_s. Each category consists of $n_1, n_2, \ldots n_s$ parameters respectively where

$$n = n_1 + n_2 + \ldots + n_s \quad (3)$$

We assume that each category comprises at most 5 parameters [1]. Figure 2 gives a pictorial view of our decomposition approach, which is based on a s-step decomposition of factorial design. As described in Sect. 1, the n parameters to be tuned are initially divided into s different categories. By performing tuning separately on each category, we effectively reduce the parameter space from $2^{n_1+n_2+\ldots+n_s}$ to $2^{n_1} + 2^{n_2} + \ldots + 2^{n_s}$ possible parameter value settings.

Our approach proceeds as follows. In the First Step, we focus on tuning the n_1 parameters, while setting the other $n - n_1$ parameters to their default values. To do that, we apply the three phases described in Sect. 2 on the training instances. At the end of this step, we obtain the best parameter setting for these n_1 parameters. Then, for the subsequent step i (from 2 to s), we tune n_i parameters by setting the best values for the parameters obtained from steps 1 to $i-1$. Once we obtain the best parameter configuration for all n parameters, we run the target algorithm with that particular configuration on the test instances.

To evaluate the performance of our proposed decomposition approach, we compare the results obtained by our approach against those obtained using the respective default parameter settings. We will show that our approach can lead to improvements in terms of the average objective values of the solutions. In the following sub-sections, we present two different decomposition approaches - one based on user guidance, and the other based on automated decomposition.

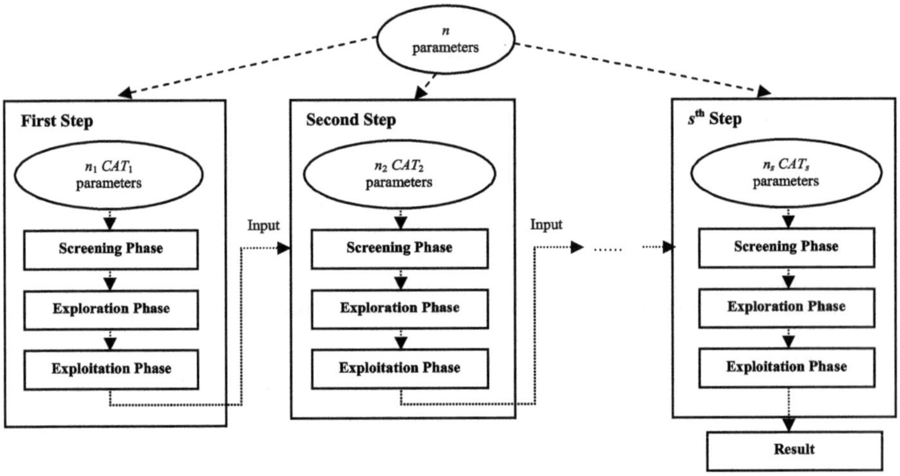

Fig. 2: s-step decomposition of factorial design

4.1 User-Guided Decomposition

In user-guided decomposition, we rely on information provided by the user for decomposing the parameters. This assumes that the user has conducted preliminary experiments a-priori and obtained observations and insights about the parameter importance levels and sensitivity to solution qualities. We then proceed to apply the s-step decomposition approach described above for each parameter category.

4.2 Automated Decomposition

A 2^k full factorial design would require a large number of runs when the number of parameters increases. An algorithm with 8 parameters would require 256 parameter settings where only 8 degrees of freedom correspond to main effects of parameters, while the rest correspond to interactions. At this point, since we have little interest in the interaction effects, we focus on decomposing the parameters based on their main effects.

For this purpose, we use a fractional factorial design to derive the parameter importance. There are several kinds of fractional factorial design classes based on existing alias relationships in the design, also known as the resolution. The concept of design resolution is to catalog fractional factorial designs according to the alias patterns they produce [15]. Resolution describes the degree to which estimated main effects are aliased (or confounded) with estimated 2-level interactions, 3-level interactions, etc. Here, we consider a Resolution IV ($2^{k-p}{}_{IV}$ design) where no main effect is aliased with any other main effect or two-parameter interactions. Note that the value of p depends on the highest possible design resolution (see Table 2) [15].

Table 2: Resolution IV design

Number of parameters (k)	p	Minimum number of runs
4	1	$2^{4-1} = 8$
5	1	$2^{5-1} = 16$
6	2	$2^{6-2} = 16$
7	3	$2^{7-3} = 16$
8	4	$2^{8-4} = 16$
9–16	Varying with k	32
17–32	Varying with k	64
33–64	Varying with k	128
...	...	...

A Resolution IV design can be used first since we assume that the parameters have monotonic effects on the response variable [15]. We take $k = 6$ parameters as an example. In a $2^{6-2}{}_{IV}$ design, only 16 parameter settings are required. To generate this design, a 2^4 design with the parameters A, B, C, D are set as the basic design, and only two additional columns are required for parameters E and F. To get the new columns, two design generators $E = A \times B \times C$ and $F = B \times C \times D$ are selected. Details of this process can be found in [15].

Note that the advantage of the above scheme is that even as the number of parameters k grows large, the minimum number of runs grows only linearly with k (rather than exponentially with the size of the original parameter space $\Theta(2^k)$). This demonstrates the scalability of our approach.

5 Experimental Results

In this section, we present the experimental results of tuning the simulated annealing algorithm (SA) applied to the spares inventory optimization problem described above. All the experiments were conducted on a PC running Windows XP with 2.33 GHz CPU and 1.96GB RAM.

50 instances were generated to represent different spares support scenarios. All instances have identical problem size, comprising three part numbers, two warehouses and three customers. A small problem size was intentionally adopted so as to allow us to have a more rigorous study of a computationally intensive spares optimization problem. Each iteration in SA involves simulating transshipment policies and inventory rationing strategies. However, the instances differ in terms of part specifications (such as unit price, failure rate, and repair turn-around-time), logistics (namely warehousing cost, transport time and cost), and contract terms (target fill rate and penalty). To decide on default parameter values, a basic inventory optimization problem, without sophisticated logistics policies or inventory rationing strategies, was used. The parameters for SA were manually tuned such that the algorithm

Table 3: Input scenarios on spares optimization problem

Specifications used to distinguish input scenarios	Range
Part specifications	
Unit price ($)	[1,000, 10,000]
Average repair cost (% of unit price)	[10, 20]
Expected repair shop turnaround time (years)	[0.01, 0.5]
Average number of spares demand at airport per year	[1. 50]
Logistics	
Average holding cost (% of unit price)	[10, 20]
Expected replenishment lead time at warehouse (years)	[0.01, 0.5]
Expected interval for transport between warehouses (h)	[1, 3]
Expected transport time between warehouses (h)	[1, 3]
Expected transport cost (% of unit price)	[10, 20]
Contract terms	
Target fill rate for operator (% of total annual demand)	[85, 100]
Delivery deadline at airport (h)	[1, 10]
Penalty rate ($)	[500, 5,000]

resulted in the best final objective value for the basic problem. The respective ranges of values are listed in Table 3.

Half of the instances were used as training instances and the rest as test instances. To account for the stochasticity involved in the SA algorithm, each instance was run 5 times and the performance of each instance was reported in terms of the average best objective value obtained and best objective value of 5 runs.

For the automated decomposition, a resolution IV fractional factorial design ($2^{8-3}{}_{IV}$ design) was adopted. The basic parameters are *Success*, *Tries*, *Reject*, *Comp* and *Oracle*. By applying principles described in [15], we define the complete alias relationships for the other three parameters: *Cooling*, *Generate* and *Change* where *Cooling* = *Success*×*Tries*×*Reject*, *Generate* = *Success*×*Tries*×*Comp* and *Change* = *Tries*×*Reject*×*Comp*×*Oracle*.

The result of the $2^{8-3}{}_{IV}$ design is summarized in Fig. 3. We observe that all parameters have p-value less than 10 %. We then proceed to classify the parameters according to the absolute main effect values. A parameter whose absolute effect value is among the top 4 would be classified as important, and less-important otherwise. Table 4 compares categories obtained by the automated-decomposition versus the user-guided one. Note that both approaches have categorized *Comp* and *Change* as CAT_1. Differences are that the user-guided decomposition categorized *Success* and *Tries* as CAT_1; and only the automated decomposition categorized *Cooling* and *Oracle* as CAT_1.

Fractional Factorial Fit: Obj versus Success, Tries, ...

```
Estimated Effects and Coefficients for Obj (coded units)

Term            Effect     Coef   SE Coef       T      P
Constant                  5.041   0.02640  190.90  0.000
Success          0.280    0.140   0.02640    5.30  0.000
Tries            0.295    0.147   0.02640    5.58  0.000
Reject          -0.152   -0.076   0.02640   -2.88  0.005
Comp            -4.146   -2.073   0.02640  -78.51  0.000
Oracle          -0.717   -0.359   0.02640  -13.58  0.000
cooling          0.619    0.310   0.02640   11.73  0.000
Generate         0.320    0.160   0.02640    6.07  0.000
Change          -3.228   -1.614   0.02640  -61.12  0.000
...........
```

Fig. 3: Statistical results of a $2^{8-3}{}_{IV}$ design

Table 4: Parameter space for SA on spares optimization problem

Parameters (p_i)	Default value	Range	User-guided	Automated	ParamILS step-size
Success	100	[100 , 1,000]	CAT_1	CAT_2	100
Tries	100	[100 , 1,000]	CAT_1	CAT_2	100
Comp	5,000	[1,000 , 5,000]	CAT_1	CAT_1	100
Reject	100	[100 , 1,000]	CAT_2	CAT_2	100
Change	2	[1 , 5]	CAT_1	CAT_1	1
Generate	300	[100 , 1,000]	CAT_2	CAT_2	100
Cooling	0.95	[0.5 , 1]	CAT_2	CAT_1	0.05
Oracle	30	[0 , 99]	CAT_2	CAT_1	10

5.1 First Step

The best values for CAT_1 parameters will be derived in this step, while setting CAT_2 parameters to their default values. Intermediate and overall results across all phases for the user-guided and automated decomposition are described below.

5.1.1 Screening Phase

Figures 4 and 5 provide results for the user-guided decomposition. Note that the dotted line in Fig. 5 represents the cut-off limit associated with the 10 % significance level. The bars at the left side of the dotted line represent insignificant parameters as well as insignificant interactions. We observe that the effect of *Comp* and *Change* are statistically significant, while both *Success* and *Tries* are insignificant. For further analysis, values for insignificant parameters can be set to a constant value. Since the spares optimization problem is a minimization problem, we set parameters with

Fractional Factorial Fit: Dev versus Success, Tries, Comp, Change

Estimated Effects and Coefficients for Dev (coded units)

Term	Effect	Coef	SE Coef	T	P
Constant		6.594	0.03561	185.18	0.000
Success	-0.061	-0.030	0.03561	-0.85	0.396
Tries	0.046	0.023	0.03561	0.64	0.521
Comp	5.162	2.581	0.03561	72.48	0.000
Change	-6.247	-3.123	0.03561	-87.71	0.000
Success*Tries	-0.002	-0.001	0.03561	-0.02	0.983
Success*Comp	-0.061	-0.030	0.03561	-0.85	0.396
Success*Change	0.132	0.066	0.03561	1.86	0.068
Tries*Comp	0.053	0.027	0.03561	0.75	0.459
Tries*Change	-0.018	-0.009	0.03561	-0.25	0.805
Comp*Change	0.518	0.259	0.03561	7.27	0.000

....

Fig. 4: Statistical results (First Step)—user-guided decomposition

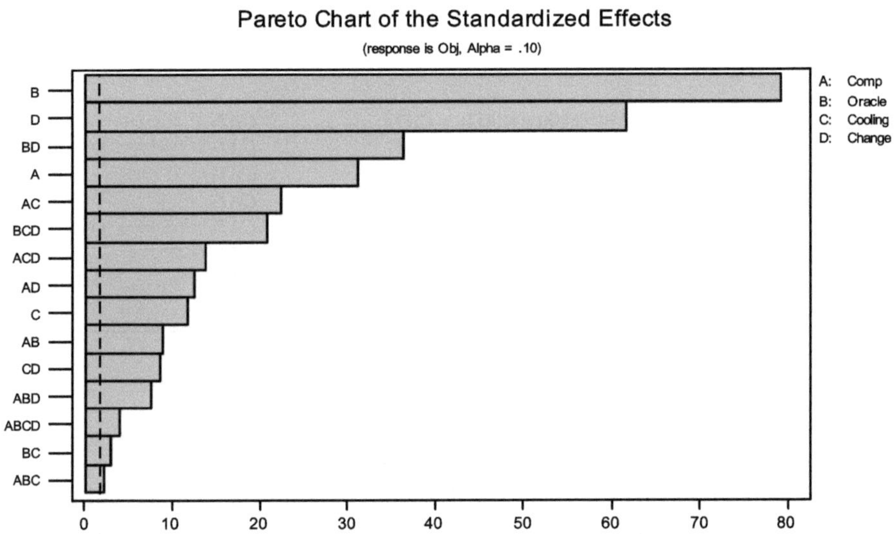

Fig. 5: Screening phase (First Step)—user-guided decomposition

negative effects to the upper-bound of the range, and those with positive effects to the lower-bound. In this case, *Success* is set to 1,000, while *Tries* is set to 100.

Figures 6 and 7 provide results for the automated decomposition. We observe that all p-values are approximately zero, implying that all important parameters (*Comp*, *Change*, *Cooling* and *Oracle*) are statistically significant. Therefore, it is not possible to define constant values for the parameters; instead, we reduce the parameter range based on the main effect value. For parameters with negative effects, the new range will span the upper half of the original range (e.g. *Comp* is adjusted from

Fractional Factorial Fit: Obj versus Comp, Oracle, Cooling, Change

Estimated Effects and Coefficients for Obj (coded units)

Term	Effect	Coef	SE Coef	T	P
Constant		4.484	0.04167	107.60	0.000
Comp	-2.599	-1.300	0.04167	-31.19	0.000
Oracle	-6.615	-3.308	0.04167	-79.37	0.000
Cooling	0.985	0.492	0.04167	11.81	0.000
Change	-5.143	-2.571	0.04167	-61.70	0.000
Comp*Oracle	0.729	0.364	0.04167	8.74	0.000
Comp*Cooling	-1.854	-0.927	0.04167	-22.25	0.000

....

Fig. 6: Statistical results (First Step)—automated decomposition

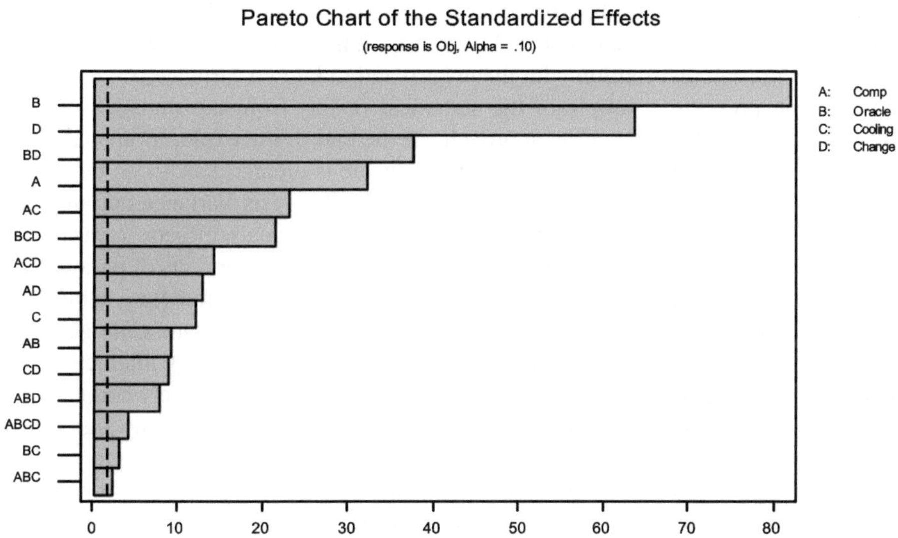

Fig. 7: Screening phase (First Step)—automated decomposition

[1,000, 5,000] to [2,500, 5,000]), and for parameters with positive effects, the new range will be limited to the lower half of the original range.

5.1.2 Exploration Phase

In this phase, we adopt the factorial experiment design to build the first-order Response Surface Model: $Y = \beta_0 + \beta_1 x_1 + \cdots + \beta_m x_m + \varepsilon$. Recall that from the screening phase, the user-decomposition approach has been simplified to consist of two parameters (*Comp* and *Change*), and four parameters (*Comp*, *Change*, *Cooling*, *Oracle*) with reduced ranges for parameter values for the automated decomposition.

Table 5: Final parameter space for SA algorithm from exploration phase (First Step)

	Final parameter space	
Parameters (p_i)	User-guided	Automated
Success	1,000	100
Tries	100	100
Comp	[1,000 , 3,000]	5,000
Reject	100	100
Change	[3 , 5]	[3 , 5]
Generate	300	300
Cooling	0.95	[0.5 , 0.75]
Oracle	30	[50 , 99]

To test the significance of the first-order model, two statistical tests, interaction and curvature tests, as described in Sect. 2, were run. It is assumed that the error component has a normal distribution with mean zero and unknown variance σ^2.

Figure 8 presents a sample of the statistical results from the statistical testing of the automated decomposition approach. At the end of the exploration phase, parameter *Comp* is no longer significant since p-value is greater than 10 %; therefore, we set its value to a high constant value. From the Analysis Variance output, each row represents the significance of each group of terms (main effects, 2-parameter interactions and so on). We observe that there are strong main effects of parameters. The row entitled "main effects" refers to the main effects of parameters, which are mainly affected by three significant parameters: *Oracle*, *Cooling* and *Change*.

The row labeled "2-Way Interactions" refers to the overall 2-way interaction between two different parameters. There may be some interaction between two parameters, such as *Oracle* and *Cooling*, *Oracle* and *Change*. There is no significant interaction between three and four parameters as well. We also conclude that the curvature test is significant. There is an evidence of curvature in the response over the region of exploration. It means that the region of the optimum has been found and then proceed to the exploitation phase to locate the best parameter setting.

In order to test the normality assumption, we generate the normal probability plot of the errors/residuals (Fig. 9). Since the errors lie approximately along a straight line, we do not suspect any problem with normality assumption. For the rest of the experiments, we conduct the same approach to test the normality of the errors.

Table 5 summarizes the final parameter space of parameters along the path of the steepest descent for user-guided decomposition and automated decomposition, respectively. This range is used as input for the exploitation phase.

5.1.3 Exploitation Phase

In this phase we apply ParamILS to tune the SA parameters based on the parameter space derived in the exploration phase. Table 6 summarizes the best parameter values obtained as a result of combining ParamILS and DOE. Since ParamILS requires

Fractional Factorial Fit: Obj versus Comp, Oracle, Cooling, ...

Estimated Effects and Coefficients for Obj (coded units)

Term	Effect	Coef	SE Coef	T	P
Constant		0.3314	0.01803	18.39	0.000
Comp	-0.0391	-0.0196	0.01803	-1.08	0.286
Oracle	-0.6303	-0.3151	0.01803	-17.48	0.000
Cooling	-0.4848	-0.2424	0.01803	-13.45	0.000
Change	-0.3617	-0.1808	0.01803	-10.03	0.000
....					
Oracle*Cooling	0.4617	0.2309	0.01803	12.81	0.000
Oracle*Change	0.3563	0.1782	0.01803	9.88	0.000
Cooling*Change	0.2615	0.1307	0.01803	7.25	0.000
....					

Analysis of Variance for Obj (coded units)

Source	DF	Seq SS	Adj SS	Adj MS	F	P
Main Effects	4	9.1755	9.17553	2.29388	147.09	0.000
2-Way Interactions	6	4.9248	4.92482	0.82080	52.63	0.000
3-Way Interactions	4	0.7313	0.73132	0.18283	11.72	0.000
4-Way Interactions	1	0.0005	0.00052	0.00052	0.03	0.856
Curvature	1	0.0842	0.08422	0.08422	5.40	0.026
Residual Error	34	0.5302	0.53024	0.01560		
Pure Error	34	0.5302	0.53024	0.01560		
Total	50	15.4467				

Fig. 8: Statistical results (First Step) for exploration phase—automated decomposition

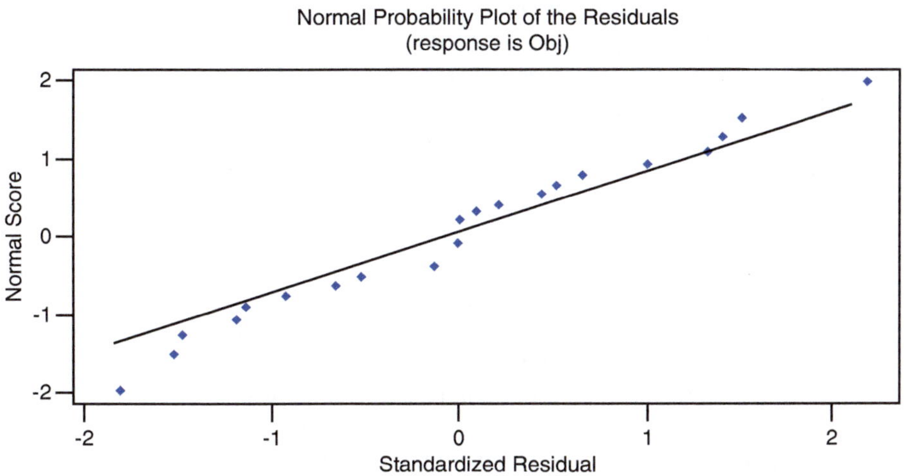

Fig. 9: Normal probability plot of errors—user-guided decomposition

all parameter domains to have discrete values, the parameters have to be discretized with a fixed step size (see last column of Table 4). Using the best parameter values, the average objective values from solving the spares optimization problem for the

Table 6: Parameter setting for SA on spares optimization problem (First Step)

Parameters (p_i)	Default value	ParamILS + DOE (user-guided decomposition)	ParamILS + DOE (automated decomposition)
Success	100	1,000	100*
Tries	100	100	100*
Comp	5,000	1,000	5,000
Reject	100	100*	100*
Change	2	3	3
Generate	300	300*	300*
Cooling	0.95	0.95*	0.6
Oracle	30	30*	90

*represents default values

25 training instances are provided in Table 7. The number of iterations of ParamILS is set to a constant number (e.g. 100). In this case, the number of iterations refers to how many times the target algorithm is being called.

We have also tried to run ParamILS on a smaller number of parameters. More precisely, we selected four parameters in CAT_1 of the user guided decomposition (Table 3), and apply ParamILS directly without exploration phase. The best total life cycle cost obtained is \$1,492,117, which is larger compared with ParamILS+DOE under the user guided decomposition (\$1,262,087—as shown in Table 6).

We observe that user-guided decomposition results in higher values for parameters *Success* and *Cooling*, while the automated decomposition results in higher values for parameters *Comp* and *Oracle*. Note that the effect of *Comp* varies in different phases of the SA algorithm. In the early phase of the search, the focus is on diversification (random walk accepting worse moves), while in the later phase, the emphasis tends towards intensification. Hence, if the parameters are tuned such that cooling is slow (i.e. high *Cooling* and high *Success*) or oracle strictness low (i.e. worse solution is more likely accepted), the range of values for *Comp* may cause the search to already terminate in the diversification phase, where there is no guarantee that a larger *Comp* value produces better solutions. On the contrary, if the SA were tuned differently such that the range of values for *Comp* extends into the intensification phase, then larger *Comp* values may probably lead to better solutions.

In our context, we observe the former case for the user-guided decomposition results from the First Step (Table 6). The solution obtained by the user-guided decomposition is 50 % more than the automated decomposition, revealing that the combination of slow cooling (0.95), lengthy exploration at each temperature (1,000 success criteria), and low oracle strictness (30), prematurely terminates the search (with maximum *Comp* value of 3,000) in the diversification phase. Coupled with the relatively large neighbourhood (change set to 3 units) for each solution, the search oscillates unpredictably and as such a *Comp* value of 3,000 for user-guided decomposition did not generate better solutions than setting *Comp* to 1,000.

Table 7: Best total life cycle cost obtained from running SA with tuned parameters (First Step) on training instances

Algorithms	Average total life-cycle cost ($)
Default value	$995,185
ParamILS + DOE - user guided decomposition	$1,262,087
ParamILS + DOE - automated decomposition	$840,921
Screening + ParamILS - user guided decomposition	$1,492,117

Fractional Factorial Fit: Obj versus Reject, Oracle, Cooling, Generate

```
Estimated Effects and Coefficients for Obj (coded units)

Term                 Effect      Coef    SE Coef       T      P
Constant                       4.8104    0.07518   63.99  0.000
Reject              -0.3067   -0.1533    0.07518   -2.04  0.046
Oracle              -0.7481   -0.3741    0.07518   -4.98  0.000
Cooling              5.4180    2.7090    0.07518   36.03  0.000
Generate             0.1938    0.0969    0.07518    1.29  0.202
Reject*Cooling      -0.0645   -0.0323    0.07518   -0.43  0.669
Reject*Generate      0.2489    0.1244    0.07518    1.66  0.103
Oracle*Cooling       0.1179    0.0590    0.07518    0.78  0.436
.......................
```

Fig. 10: Statistical results (Second Step)—user-guided decomposition

5.2 Second Step

In this step, values for CAT_2 parameters as listed in Table 4 will be determined. The values of CAT_1 parameters are set to the best values obtained by ParamILS + DOE in Step 1 (Table 6).

5.2.1 Screening Phase

Figure 10 presents the results of CAT_2 parameters for the user-guided decomposition approach. All parameters except parameter *Generate* are found to be significant. Since *Generate* has a positive main effect, it will be set to the lower bound of the original range. The intervals for the statistically significant parameters are reduced such that parameters with negative effects are assigned a new range covering the upper half of the original range (e.g. *Oracle* is adjusted to [50, 99]), and parameters with positive effects will be limited to the lower half of the original range.

Similar observations can be obtained from the results of the automated decomposition (Fig. 11). Only parameters *Success* and *Tries* are statistically significant. The other two parameters, *Reject* and *Generate*, are set to their lower bound. In

Fractional Factorial Fit: Obj versus Success, Tries, ...

```
Estimated Effects and Coefficients for Obj (coded units)

Term                    Effect      Coef   SE Coef      T      P
Constant                        0.332437   0.01214  27.38  0.000
Success               0.541432  0.270716   0.01214  22.29  0.000
Tries                 0.433361  0.216680   0.01214  17.84  0.000
Reject                0.013859  0.006929   0.01214   0.57  0.572
Generate              0.029397  0.014698   0.01214   1.21  0.235
Success*Tries         0.556309  0.278155   0.01214  22.91  0.000
Success*Reject        0.011582  0.005791   0.01214   0.48  0.637
Success*Generate      0.029080  0.014540   0.01214   1.20  0.240
.....................
```

Fig. 11: Statistical results (Second Step)—automated decomposition

Table 8: Final parameter space for SA algorithm from exploration phase (Second Step)

	Final parameter space	
Parameters (p_i)	User-guided	Automated
Success	1,000	[100 , 500]
Tries	100	[100 , 500]
Comp	1,000	5,000
Reject	800	100
Change	3	3
Generate	100	100
Cooling	[0.5 , 0.6]	0.6
Oracle	[80 , 99]	90

the next phase, we only focus on parameters *Success* and *Tries* for the automatic decomposition.

5.2.2 Exploration Phase

In this phase, we apply the same approach as we have used in the First Step until we reach the region of the optimum. The final parameter space of parameters for both decomposition approaches are summarized in Table 8. For the user-guided decomposition, parameter *Reject* was found to be insignificant at the end of the exploration phase; hence it is set to a constant value equal to the lower bound from the screening phase.

Table 9: Parameter setting for SA on spares optimization problem (Second Step)

Parameters (p_i)	Default value	ParamILS + DOE (user-guided decomposition)	ParamILS + DOE (automated decomposition)
Success	100	1,000	500
Tries	100	100	300
Comp	5,000	1,000	5,000
Reject	100	800	100
Change	2	3	3
Generate	300	100	100
Cooling	0.95	0.55	0.6
Oracle	30	80	90

5.2.3 Exploitation Phase

Table 9 summarizes the best parameter values obtained as a result of combining ParamILS and DOE for the Second Step. As in the First Step, the three sets of parameter values were used for solving the training instances with SA. We conducted 5 runs on each instance and the average as well as the best found objective value for each instance is determined (Table 10). For each instance, the percentage improvement of each proposed decomposition approach from the default case (denoted as % improvement over average and best results, respectively) are calculated. We observe that the best objective values obtained in the Second Step outperform those of the First Step. In addition, by applying ParamILS+DOE, further improvements can be obtained for both training and test instances. The grand mean of average total life cycle costs of our proposed approaches are statistically different from those of default value setting (with the 10 % significance level).

In general, it is found that automated decomposition consistently yields better solutions than user-guided decomposition in terms of the grand mean of the average and best total life cycle cost. All training and test instances solved by the automated decomposition outperform the default case, while the user-guided decomposition provided better solutions for 18 training and 20 test instances only. Furthermore, the percentage improvement by using the automated decomposition is higher, achieving improvements of the best values of 16.28 % and 14.30 % for training and test instances, respectively, while the user-guided cases only achieved 1.33 % and 3.27 % respectively. Similar results are obtained for improvements in average objective values—the automated decomposition achieved 12.88 % and 12.01 % improvements in training and test instances, respectively, compared with the user-guided decomposition achieving 1.81 % and 4.53 % respectively. Operationally, improvements achieved by the automated-decomposition amounts to an average total life cycle cost savings of more than 10 %, which in our test cases is worth $100,000 as compared to the default case.

The standard deviation (stdev) and average of coefficient of variance (CV) reaffirms the superior performance of the automated decomposition. As seen from Ta-

Table 10: Parameter tuning and results for SA on spares optimization problem (Second Step)

Results	Default value	ParamILS + DOE (user-guided decomp.)	ParamILS + DOE (automated decomp.)
Training instances			
Grand mean of average total life cycle cost ($)	995,608.80	969,816.90	837,820.40
p-value	–	0.079	0
N instances with better results	–	18	25
N instances with worse results	–	7	0
% improvement over average results	–	1.33 %	16.28 %
Average of best total life cycle cost ($)	956,164.70	929,390.30	837,109.00
p-value	–	0.025	0
N instances with better results	–	17	25
N instances with worse results	–	8	0
% improvement over best results	–	1.81 %	12.88 %
Average of std dev total life cycle cost	38,869.3	33,089.2	570.6
Average of CV total life cycle cost	3.86 %	3.63 %	0.00 %
Test instances			
Grand mean of average total life cycle cost ($)	1,169,127.00	1,125,680.00	1,004,810.00
p-value	–	0	0.001
N instances with better results	–	20	25
N instances with worse results	–	5	0
% improvement over average results	–	3.27 %	14.30 %
Average of best total life cycle cost ($)	1,137,053.00	1,082,371.00	1,002,520.00
p-value	–	0	0
N instances with better results	–	20	25
N instances with worse results	–	5	0
% improvement over best results	–	4.53 %	12.01 %
Average of std dev total life cycle cost	33,072.7	39,142	2,369.3
Average of CV total life cycle cost	2.90 %	3.62 %	0.19 %
Execution time	14 h	10 h	15 h

ble 10, stdev for the automated decomposition are relatively low for both training and test instances (at 570.6 and 2,369.3, respectively). A low standard deviation indicates that the results of several runs tend to be consistent and close to the average total life cycle cost. We also calculated the average of CV in order to measure the dispersion of results against the average total life cycle cost. All three approaches achieved low values of CV (less than 5 %). The automated decomposition is the lowest (best) compared with the default case and the user-guided decomposition approaches. In terms of average execution time for solving an instance, the automated decomposition requires 1 h and 5 h more than the default case and user-guided decomposition, respectively.

Table 11: ParamILS versus ParamILS + DOE (automated decomposition)

Results ($)	ParamILS	ParamILS+DOE (automated)	% Improvement
Training instances			
Grand mean of average total life cycle cost($)	1,007,763	837,820	16.86
Average of best total life cycle cost($)	961,296	837,109	12.92
Test instances			
Grand mean of average total life cycle cost($)	1,236,578	1,004,810	18.74
Average of best total life cycle cost($)	1,198,951	1,002,520	16.38

Finally, we compare the results obtained by the automated decomposition with the pure application of ParamILS (i.e. without the decomposition of parameters). In this case, the computational budget allocated (i.e. the number of iterations) is fixed. For instance, suppose the numbers of iterations of ParamILS and DOE are x and y respectively, the number of iterations of the first scenario is $x+y$, while the number of iterations of the second scenario is set to z, with $z = x+y$. Here, the number of iterations of ParamILS and DOE are 100 (as mentioned in Sect. 5.1.3) and $(5 \times 2^{n-3} + 5 \times 2^{n_1} + 5 \times 2^{n_2})$, respectively; while the number of iterations of the pure ParamILS is the total of both number of iterations.

Table 11 shows that for both training and test instances, the combination of ParamILS and DOE outperforms that of a direct application of ParamILS. We observe an improvement on the average of best total life cycle cost of 12.92 % and 16.38 % for training and test instances respectively.

6 Conclusion

In this paper, we explore the idea of improving the efficiency of factorial design for parameter tuning of metaheuristics. Our experimental results show that cost savings exceeding 10 % can be achieved. Due to the non-convexity of the search space, the solution converges to a local optimum when SA was manually configured. The tuned SA using our approach resulted in the significant annual cost savings because arbitrage opportunities, otherwise missed in the default SA settings, were identified.

For future extensions to this work, one can consider using other approaches such as Cluster Analysis to classify (decompose) parameters. Cluster analysis is a statistical approach used to divide the parameters into a set of objects based on their similarity (see [9]). The basic assumption in the first order (planar) model is that errors are assumed to be uncorrelated and distributed with mean 0 and constant (but unknown) variance. A normal plot of the errors should be approximately a straight line. Transformation of the response would be necessary if the normal plot is not a straight line [16]. We will also consider a more elaborate model such as the second order model in order to find the optimum region since interaction and curvature do exist [15].

In this paper, we make use of one automated parameter tuning configurator, namely ParamILS, and apply a 2-step (2 categories) decomposition approach. One can consider using other configurators, such as F-Race, for solving other real-world parameter tuning problems with larger number of parameters. Such a case study would also be interesting to explore a more general s-step decomposition.

Acknowledgements The authors wish to thank Lindawati and Nguyen Van Vinh for their valuable assistance in the technical implementation of the algorithm.

References

1. Adenso-Díaz, B., Laguna, M.: Fine-tuning of algorithms using fractional experimental design and local search. Oper. Res. **54**, 99–114 (2006)
2. Ansótegui, C., Sellmann, M., Tierney, K.: A gender-based genetic algorithm for the automatic configuration of solvers. In: Proceedings of the 15th International Conference on Principles and Practice of Constraint Programming, (CP 2009). Lecture Notes in Computer Science, vol. 5732, pp. 142–157. Springer, Heidelberg (2009)
3. Axsáter, S.: Modelling emergency lateral transshipments in inventory systems. Manag. Sci. **36**, 1329–1338 (1990)
4. Birattari, M., Stützle, T., Paquete, L., Varrentrapp, K.: A racing algorithm for configuring metaheuristics. In: Proceedings of the Genetic and Evolutionary Computation Conference (GECCO 2002), pp. 11–18. Morgan Kaufmann, San Francisco (2002)
5. Birattari, M., Yuan, Z., Balaprakash, P., Stützle, T.: F-race and iterated F-race: An overview. In: Bartz-Beislstein, T., Chiarandini, M., Paquete, L., Preuss, M. (eds.) Empirical Methods for the Analysis of Optimization Algorithms, pp. 311–336. Springer, Berlin (2010)
6. Cattani, K.D., Souza, G.C.: Inventory rationing and shipment flexibility alternatives for direct market firms. Prod. Oper. Manag. **11**, 441–457 (2002)
7. Coy, S.P., Golden, B.L., Runger, G.C., Wasil, E.A.: Using experimental design to find effective parameter settings for heuristics. J. Heuristics **7**, 77–97 (2000)
8. Gunawan, A., Lau, H.C., Lindawati: Fine-tuning algorithm parameters using the design of experiments approach. In: Coello, C.A.C. (ed.) Proceedings of Learning and Intelligent Optimization, 5th International Conference (LION 5). Lecture Notes in Computer Science, vol. 6683, pp. 278–292. Springer, Heidelberg (2011)
9. Han, J., Camber, M.: Data Mining: Concepts and Techniques, 2nd edn. Morgan Kaufmann, San Francisco (2006)
10. Hutter, F., Hoos, H.H., Leyton-Brown, K., Stützle, T.: ParamILS: an automatic algorithm configuration framework. J. Artif. Intell. Res. **36**, 267–306 (2009)
11. Hutter, F., Hoos, H.H., Leyton-Brown, K.: Automated configuration of mixed integer programming solvers. In: Lodi, A., Milano, M., Toth, P. (eds.) Proceedings of International Conference on Integration of Artificial Intelligence and Operations Research techniques in Constraint Programming (CPAIOR). Lecture Notes in Computer Science, vol. 6140, pp. 186–202. Springer, Heidelberg (2010)
12. Hutter, F., Hoos, H.H., Leyton-Brown, K., Murphy, K.P.: Time-bounded sequential parameter optimization. In: Blum, C., Battiti, R. (eds.) Proceedings of Learning and Intelligent Optimization, 4th International Conference (LION 4). Lecture Notes in Computer Science, vol. 6073, pp. 281–298. Springer, Heidelberg (2010)

13. Hutter, F., Hoos, H.H., Leyton-Brown, K.: Sequential model-based optimization for general algorithm configuration. In: Proceedings of Learning and Intelligent Optimization, 5th International Conference (LION 5). Lecture Notes in Computer Science, vol. 6683, pp. 507–523. Springer, Heidelberg (2011)
14. Lau, H.C., Xiao, F.: Enhancing the speed and accuracy of automated parameter tuning in heuristic design. In: Proceedings of the 8th Metaheuristics International Conference (MIC 2009), Hamburg, Germany (2009)
15. Montgomery, D.C.: Design and Analysis of Experiments, 6th edn. Wiley, New York (2005)
16. Ridge, E., Kudenko, D.: Sequential experiment designs for screening and tuning parameters of stochastic heuristics. In: Proceedings of Workshop on Empirical Methods for the Analysis of Algorithms, pp. 27–34. Reykjavik, Iceland (2006)

Evolving Pacing Strategies for Team Pursuit Track Cycling

Markus Wagner, Jareth Day, Diora Jordan, Trent Kroeger, and Frank Neumann

Abstract Team pursuit track cycling is a bicycle racing sport held on velodromes and it is part of the Summer Olympics. It involves the use of strategies to minimize the overall time that a team of cyclists needs to complete a race. We present an optimisation framework for team pursuit track cycling and show how to evolve strategies using metaheuristics for this interesting real-world problem. Our experimental results show that these heuristics lead to significantly better strategies than state-of-art strategies that are currently used by teams of cyclists.

1 Introduction

Metaheuristics such as local search, simulated annealing, evolutionary algorithms, and ant colony optimisation have been shown to be very successful problem solvers in various domains [8]. This paper describes a novel application of metaheuristics to the optimisation of elite competitive track cycling. Specifically, we examine how they can be used to optimise strategies that lead to improved performance in the women's team pursuit event.

Mathematical models of cycling performance have been proposed that aim to predict the outcome of elite single cyclist track and road races [6, 5]. These models balance the estimated anaerobic and aerobic power capabilities of a cyclist against resistive forces such as aerodynamic drag and surface friction in order to determine a velocity profile that corresponds to maximal effort for the cyclist. However, given the non-linear nature of these models, an all-out maximal effort may not necessarily lead to optimal performance. To address this, mathematical optimisation techniques have been applied to the problem of determining an optimal pacing strategy for a single cyclist in track sprint events [2]. The work described in this paper builds

Markus Wagner · Jareth Day · Diora Jordan · Trent Kroeger · Frank Neumann
School of Computer Science, The University of Adelaide, Adelaide, SA, Australia
e-mail: markus.wagner, jareth.day, diora.jordan, trent.kroeger, frank.neumann@adelaide.edu.au

L. Di Gaspero et al. (eds.), *Advances in Metaheuristics*, Operations Research/Computer Science Interfaces Series 53, DOI 10.1007/978-1-4614-6322-1_4,

upon this previous research to propose a framework capable of optimising pacing strategies for a team of cyclists, a problem context in which significantly more complex dynamics are possible.

We present a problem formulation for the team cycling problem and show how to put it into an optimisation framework. The framework takes into account that cyclists upfront spend more energy than cyclists that are in the slipstream of others. This implies that cyclists have to change their positions within the team during the race in order to minimize the overall time. In our optimisation problem, we have to determine when such changes occur as well as the power applied by the whole team during different parts of the race.

Our problem has some relation to scheduling problems. Scheduling problems have been addressed in operations research and computer science for several decades. Recent examples of real-world scenarios where evolutionary techniques were applied are, amongst others, vehicle scheduling in public transport [1]. Our problem has similarities to the scheduling of jobs with controllable processing times [7] as we have to determine at each point of time the power that the team applies in the race. However, it involves more difficulties as we also have to determine a strategy for changing the order of the riders.

From an algorithmic perspective, the problem is interesting and challenging as it requires the simultaneous optimisation of discrete and continuous variables. The discrete variables determine the scheduling of the different cyclists in the lead position during a race, whereas the continuous variables determine the level of power applied by that cyclist. We present different approaches for dealing with this mixed discrete and continuous setting and evaluate the quality of the different approaches with respect to standard strategy and power settings. Our results show that state-of-the art strategies for team pursuit track cycling can be significantly improved using metaheuristic approaches.

This paper is structured as follows. Section 2 outlines the specific problem to be solved in detail, with its formulation as an optimisation problem following in Sect. 3. In the subsequent Sect. 4, our evolutionary approach to this problem is described, with results being presented and discussed in Sect. 5. The paper concludes in Sect. 6 with a summary of key findings and a description of potential areas for future research.

2 Team Pursuit Track Cycling

Track cycling is raced on elliptical tracks called velodromes. These can be indoor or outdoor and are usually 250 m in circumference, which includes two straight sections and two banked turns. Track cycling comprises a number of sprint and track endurance events. Both individual and team events exist. These include pure sprint events such as the individual and team sprints; long sprints such as the 500 m and 1,000 m time trials and the *Kieren*; middle distance events such as the individual and team pursuits; and endurance events such as the *Madison*, scratch, handicap and points races. As described in the previous section, mathematical models of performance have been developed for predicting and optimising a subset of the individual events. This paper focuses on the application of a specific metaheuristic to optimise both the pacing strategy, in terms of the power applied by each cyclist in first position, and the transition strategy employed within the context of a women's team pursuit event.

The team pursuit is an event in which multiple cyclists work together to complete a given race distance in the minimum possible time. The term "pursuit" is derived from the fact that races begin with two opposing teams that begin on opposite sides of the track and try their best to pursue each other. Team members take turns riding in the front position thus allowing the other team members to draft closely behind for maximum aerodynamic benefit. This rotation allows the team to maintain a significantly higher velocity than would be achievable by a single cyclist. Changes to the relative position of the cyclists are most efficiently performed on either of the two banked turns of the track, where the cyclist in first position sweeps up the bank of the turn and then rejoins in last position. For the men's team pursuit event, there are four cyclists in each team and the race consists of sixteen laps for a total of 4,000 m. In this event, only three of the four cyclists need to complete the race and so it is common to see one cyclist sacrifice themselves for the team by spending longer than average in first position to the point where his energy reserves are exhausted and he leaves the race. In the women's team pursuit event, there are three cyclists in each team and the race is over twelve laps for a total of 3,000 m. Unlike the men's variant, all three cyclists in the women's team pursuit must complete the entire race distance, which typically leads to a more even distribution of workload amongst the team. The standard transition strategy observed from world championship and Olympic women's team pursuit events usually involves cyclist transitions occurring after 187.5 m (0.75 of a lap) and then every 250 m (1 lap) until the 2,562.5 m point (1.25 laps from the finish), after which no further transitions occur.

Due to its lower complexity, the women's team pursuit event was chosen as the basis for the work described in this paper. There are four main ways by which to improve performance in the women's team pursuit event. The first is to improve the physiological and psychological capabilities of the cyclists through training. The second is to improve the technical specifications of the bicycles used, in terms of their mass and aerodynamic properties. The third is to change the pacing strategy used by the team such that the power applied results in maximal benefit. Finally, the transition strategy may be changed to achieve a more even distribution of energy usage by the team and the most aerodynamically effective velocity profile. This paper will focus on the application of metaheuristic algorithms to optimise both the pacing strategy, in terms of the power applied by each cyclist in first position, and the transition strategy employed within the context of a women's team pursuit event. This problem is interesting in that it requires the simultaneous optimisation of both discrete and continuous variables. The discrete variables in this problem identify the specific positions within the race where a transition occurs. These transition points are constrained to occur only on the two banked turns of the track and may therefore be represented by integer variables that identify half-laps throughout the race. The continuous variables describe the power to be applied by the cyclist in first position between each transition, which may vary continuously between zero and the maximum power within the physiological capability of the cyclist. The non-linear nature of the physiological and kinematic cycling model used for this project means that it is difficult, if not impossible, to frame this problem in terms of traditional mathematical optimisation approaches and so we have developed a hybrid metaheuristic optimisation framework that simultaneously optimises these discrete and continuous variables.

3 Problem Formulation

In Sect. 2, we discussed the nature of team pursuit cycling races, and noted that the aim of our optimisation was the minimisation of the race time. We determined that the core parameters to be optimised are the number of half-laps the first rider performs before the cyclists change position and the power output of the first rider. We thus have two sets of parameters that we need to form a solution for the race model, the half-laps before rider transition and the first rider power output per lap.

A transition moves the first cyclist to the rear and the second to the front. These transitions occur exclusively on one of the two banked turns and so there is a minimum of one half-lap between transitions. The power output of the first rider per half-lap is able to be simplified in that a rider should maintain a certain power level at all times while in the lead. Due to this, only one power output value is needed per half-lap before transition. We aim to maintain a fixed number of variables for simplicity; with the worst case of riders transitioning every half-lap we find that it is necessary to store a number of variables equal to twice the number of half-laps being represented.

While the half-laps before transition can simply be represented by integers, the power output is measured in W and must be a floating point number to correctly capture precision. Due to both integer and floating-point values represented in the solution, we choose a floating-point-valued representation in order to maintain a simple, fixed-length solution without sacrificing precision in the manner discussed by Herrera et al. [4].

In Sect. 2 we noted that there is typically an initial 0.75 laps, or 1.5 half-laps, required at the start of the race before transitions can be made, and a final 1.5 half-laps at the end of the race in which transitions may not be made. In this way, these initial and final race segments are atomic and are hence represented within the solution in the same manner as regular half-laps. Rather than representing n half-laps, we instead represent $n-2(1.5)+2$. As we are storing two sets of information based on the number of transitions set to the maximum of $n-1$, the total number of parameters in the representation is $2(n-1)$. Figure 1 illustrates the solution representation.

If any half-laps before transition is greater than one, there will be extraneous information in both halves of the solution. We are thus interested only in the first m variables on both halves of the solution, where m is derived from $(\sum_{k=1}^{m} HL_k) \geq$

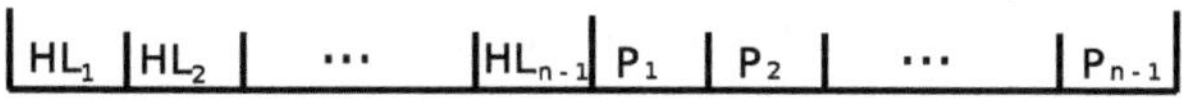

Fig. 1: Illustration of the solution representation. *HL* denotes the half-laps before rider transition and *P* indicates power output per first rider per half-lap

$(n-1)$ and HL_k is the kth element of the half-laps before transition set. To illustrate, given a solution with the half-laps before transition set equal to [3, 3, 3, 3, 3, 3, 3, 3, 3, 3, 3, ...] and $n = 24$, we would only be interested in the first eight half-laps before transition and the corresponding eight power levels since $(\sum_{k=1}^{8} HL_k) = 24 \geq 23$.

For example, if we allow values of 1...3 for each transition position, a total of 755,476 transition strategies is possible. A range of 1...4 results in 2,033,628 different combinations, and a range of 1...5 in 3,040,048 different combinations.

3.1 Team Pursuit as an Optimisation Problem

The fitness evaluation function uses time as a measure of fitness while making sure that the energy expended for each rider throughout the race is within the physiological capability of that rider. In order to calculate the given time for a race we perform two tasks. First, we modify the model taken from [5] in order to be suitable for the given problem. This then provides the formulae to calculate the power for the riders to ride a given distance. Second, we use forward integration modelling in order to predict the time taken using the solution being evaluated. Both tasks are set into context in pseudo-code form in Algorithm 1, and are explained in greater detail in the following.

Algorithm 1 Fitness Function: Evaluating a Pacing Strategy

Input: list of transitions *transitions*, list of associated power levels *powers*
Output: the predicted race time *totalTime*

set $rider_1$, $rider_2$, and $rider_3$ according to their preset initial order
$totalTime := 0$
for $i \in \{1, \ldots, s.length\}$ **do** {iterate in order over the transition strategies}
    $h := transitions[i]$ {get the current number of half laps}
    $d := h \cdot halfLapLength$ {set the current distance rode}
    $p := powers[i]$ {get the current power level}
    $t := forwardIntegration(\ldots)$ {compute the time needed for cycling d, using Eq. (2)}
    $totalTime \mathrel{+}= t$ {increase the race time by the time $rider_1$ needs to cycle the distance d}
    $rider_1.totalEnergy \mathrel{-}= p \cdot t$ {reduce the first rider's energy}
    $useEnergy(rider_2)$ & $useEnergy(rider_3)$ {reduce the other riders' energies using Eq. (1), based on them keeping up with $rider_1$'s speed}
    $swapRiderConfig()$ {reorder the riders}
    $totalTime \mathrel{+}= transitionTime$ {increase the race time by the time lost by reordering the riders}

3.1.1 Task 1: Modified Model for Team Pursuit

The model taken from [5] was adapted to suit track cycling conditions. This involved removing the potential energy as vertical elevation is negligible, and combining air speed and ground speed as the cycling is performed in still air conditions. This gives us the following formula:

$$\Delta KE = (P \times E - C_D A \times (\frac{1}{2}\rho \mathrm{v}^3) - \mu \times (\mathrm{v}F_N)) \times \Delta t \quad (1)$$

in which KE is kinetic energy, P is power, E is mechanical efficiency, C_DA is the frontal area of the bike, wheel spokes and rider, ρ is air density, v is speed, μ is a global coefficient of friction, F_N is the weight of the bike and the rider, and t is time. We then calculate the acceleration over a given distance, which is obtained from the number of laps in a given configuration for the current solution, by assuming that the excess kinetic energy is used for this acceleration. The acceleration is then used to calculate the time taken for the rider to cover the given distance.

As the second and third rider have a drafting benefit from being in the slipstream of the first rider, we use the following formula[1] to calculate the power used to keep up with the acceleration of the first rider:

$$P = (C_D A \times C_{Draft} \times \frac{1}{2}\rho \mathrm{v}^3 + \mu \times (\mathrm{v}F_N) + \frac{\Delta KE}{\Delta t})/E \quad (2)$$

in which E is the efficiency of the drive system, and where ΔKE is calculated using the final velocity of the first rider after acceleration. C_{Draft} is drafting coefficient of the rider and represents the reduction in the C_DA of a cyclist due to the aerodynamic benefits of drafting in second or third position.

3.1.2 Task 2: Forward Integration Modelling for Race Time Computation

The algorithm steps through the different configurations the cyclists would transition to throughout the race. When the riders transition, it is assumed to take a fixed amount of time which is added to the total time of the race. For each of the different configurations, the number of half laps rode in the current configuration is used to calculate the distance rode. The time taken to ride this distance is then calculated by forward integration, using Eq. (1), and a small Δt (0.1 s). This time is then added to the total race time. The energy used to ride the distance in this configuration is subtracted from the total energy of the first rider, in order to ensure the rider does not run out of energy. This total energy is obtained from the initial amount of available energy each rider is assumed to have, which can be arbitrarily allocated throughout the race within physical limitations. Equation (2) is then used to calculate the energy used by the second and third rider in order to maintain pace with the first rider.

[1] It is an extension based on [5] to incorporate drafting benefits.

When the cumulative sums of the distances cover the race distance, the time taken is then used as a measure of the fitness of the solution currently being evaluated.

4 Metaheuristic Approach

When optimising the overall race strategies, we faced the problem of optimising both the transition strategy and the power profile at the same time. By splitting up the problem into two separate ones, namely (1) the continuous power optimisation and (2) the discrete strategy optimisation, this allowed us to significantly reduce the search space.

4.1 Power Profile Optimisation

With the use of a powerful optimiser for the first part, we could concentrate our efforts on finding transition strategies. The evaluation of a single race strategy is the race time of its transition strategy, for which an optimised power profile is computed with respect to the given fitness function.

We use the state-of-the-art evolutionary algorithm called Covariance Matrix Adaptation Evolution Strategy (CMA-ES) [3] as our local optimiser for the power profiles. It self-adapts the covariance matrix of a multivariate normal distribution. This normal distribution is then used to sample from the multidimensional search space where each variate is a search variable. The covariance matrix allows the algorithm to respect the correlations between the variables making it a powerful evolutionary search algorithm. Its internal management of perturbation range and other parameters allows it to overcome problems such as parameter tuning, poor performance on badly scaled fitness functions, the need to scale population sizes in proportion to problem dimension, and premature convergence of the population.

In our setup, the initial population of CMA-ES's optimisation process is set close to the power levels of the currently best solution x^*, with the initial standard deviation being $\pm 10\,\mathrm{W}$. This way, time is saved to arrive in the right order of magnitude regarding the power levels, compared to starting with randomly initialized solution vectors.

During the power level optimisation, constraint violations are handled by the fitness function, which adds high weights to the predicted race time in case a rider runs out of energy or the power levels exceed the limits.

Later on, in Sect. 5.2, we will briefly compare the performance of CMA-ES as our local optimiser to a simple random search variant, where the latter turns out to be inferior and even counterproductive regarding the resulting race time.[2]

[2]Note that other optimiziation software may outperform CMA-ES, but as its algorithmic setup is virtually parameter-free, no fine-tuning of parameters was required.

4.2 Transition Strategy Optimisation

When we constructed our algorithms for the strategy optimisation, we had to consider that the above-described continuous power profile optimisation is relatively time-consuming, with each evaluation being a whole run of CMA-ES, and each taking several seconds. We therefore chose to use random local search (RLS) and a simple evolutionary algorithm (Simple EA) variant, as described (see Fig. 2). Both use only a population size of $\mu = 1$, and we conjecture that larger population sizes are not necessarily beneficial in the face of the significantly increased runtime of the algorithm.

For the initial transition strategies, we randomly select for each position in the strategy a number from 1 to 3.

Note that our applied operators first determine the effective length m of a strategy, i.e. the subsequence (starting at the first entry) that is sufficient to finish the race (as described in Sect. 3).

For RLS, the neighbourhood is defined as follows: given the current transition strategy s, the neighbourhood contains all those strategies s' that can be reached by incrementing (respectively decrementing) a single field in the strategy. Note that this can potentially change the strategy's effective length. However, this is unproblematic because the individual's length is sufficient to guarantee a valid strategy to complete the race.

The mutation operator used in the Simple EA randomly changes the value of a field of the effective strategy to a different value. Whenever a new individual is created, the expected number of such changes is one, as the probability of performing such a change is $p = 1/m$. Thus, contrary to RLS, several positions can be modified during one generation of an offspring.

Algorithm 2 Random Local Search

```
Choose x* uniformly at random.
repeat forever
    repeat
        if there is an unseen neighbour of x* then
            Create said neighbour x of x*.
        else
            End of algorithm.
    until f(x) ≤ f(x*)
    x* := x
```

Algorithm 3 Simple EA

```
Choose x* uniformly at random.
repeat
    Create x by mutating each position of x* independently with probability 1/m.
    if f(x) ≤ f(x*) then
        x* := x
until maximum number of fitness evaluations reached.
```

Fig. 2: The algorithms used for the transition strategy optimisation. The fitness function f evaluates a given strategy x by finding an optimised power profile for it, and then returning the race time based on x and the profile

Table 1: Race characteristics

Gravitational acceleration	9.80665 ms^{-2}	Rider A mass	70.0 kg
Mechanical efficiency	0.977	Rider B mass	67.5 kg
Global friction	0.0025	Rider C mass	65.0 kg
Temperature	20° C	Available energy	mass × 5 × 210 J
Air pressure	1,013.25 hPa	Bicycle mass	8.0 kg
Relative humidity efficiency	50 %	Rider A C_DA	0.190
Race distance	3,000 m	Rider B C_DA	0.175
Maximum power	1,000 W	Rider C C_DA	0.160
Minimum power	100 W	C_{Draft} second position	0.7
Maximum half-laps before transition	3	C_{Draft} third position	0.6
Laps	12	Time to transition	0.12 s

5 Experiments

5.1 Experimental Parameters

To make a realistic model of the race, we have implemented a number of parameters as listed in Table 1.[3] Of note, time to transition is the time lost when the riders transition. We approximate the power a rider can output by their personal mass multiplied by 5 Wm^{-1}, and determine that the available energy is this power over 210 s. We use a standard 8 kg for a bicycle's mass. C_DA is the effective frontal area of the rider. Finally, we impose an upper bound of 1,000 W and a lower bound of 100 W to ensure a feasible solution, and model fatigue by implementing a maximum number of half-laps before a rider has to transition.[4]

5.2 Power Profile Optimisation

In order to demonstrate the fitness evaluation, and to form a basis for comparison with optimised solutions, we investigate a set transition strategy: the first rider performs the initial 0.75 laps, and then the riders transition every two half-laps. This transition strategy is represented as [1, 2, 2, 2, 2, 2, 2, 2, 2, 2, 2, 2]. We use the initial rider configuration [A, B, C], allowing the riders with the least energy to gain benefit from drafting. To increase readability, we use the simpler notation of three capital letters instead of the list style.

We investigate two different power profiles for this demonstration. The first profile, Unoptimised 1, has the first rider output a greater amount of power in order to

[3]Based on private communication with the Australian Institute of Sport cycling team.

[4]Note that other ways to model fatigue can be incorporated into the fitness function, such as a declining maximum performance if the first rider stays in front for too long.

Table 2: Resulting times on unoptimised and iterative solutions with the initial rider configuration ABC

Power profile	Race time (s)	Rider A final energy (J)	Rider B final energy (J)	Rider C final energy (J)
Unoptimised 1	208.42	39.89	1294.01	1861.89
Unoptimised 2	209.92	21.50	549.97	1657.38
Iterative 1	208.92	57.06	30.10	65.18
Iterative 2	213.22	28.57	26.67	48.36

reach a higher velocity by the end of the initial 0.75 laps, but upon transition has the riders drop back to a lower power for the rest of the race. We use the arbitrarily high value of 900 W with a lower constant of 364 W, the highest that can be used while maintaining positive rider energy.

The second profile, Unoptimised 2, uses a constant power throughout the race, sacrificing initial velocity for a smoother acceleration. We use 409 W, the highest constant power value possible without entering negative energy values by the end of the race. Table 2 illustrates the race time and remaining energy for the two solutions with the initial rider configuration ABC.

In addition, we created an algorithm that generates a solution iteratively by cycling through the same transition strategy, attempting to add or subtract power from the power profile in order to minimize the remaining energy of all the riders at the end of the race. We again use the initial rider configuration ABC, and create two different power profiles. The first solution, Iterative 1, aims for a high energy output for the first 0.75 laps, and the second, Iterative 2, aims for an even power distribution. Table 2 shows the race time for the resulting solutions. The first iterative solution has the power profile [924, 350, 393, 324, 350, 393, 324, 350, 393, 324, 350, 393], and the second iterative solution has the power profile [373, 355, 460, 373, 355, 460, 373, 355, 460, 373, 355, 460].

We can see from Table 2 that iteratively altering the power to maximize rider power output will not necessarily return a better solution. Although the riders have less energy at the end of the race in the iterative solutions, the best iterative solution is slower than the best unoptimised solution.

We next run this set transition strategy under CMA-ES in order to determine if we can gain a faster race time by optimising the power profile. We use CMA-ES on this single solution for 100 repetitions, each with 2000 evaluations. Table 3 lists the results for the optimised strategy.

This optimised solution uses the same transition strategy with the power profile [759, 430, 410, 464, 430, 398, 466, 432, 396, 294, 328, 298]. It returns a best time of 204.52, 3.90 s better than the best unoptimised time. Figures 3 and 4 compare the power and velocity profiles of the optimised and the two unoptimised solutions.

In addition to the initial ABC rider configuration, we are able to extend the model by investigating the optimisation of all permutations of the riders. Table 4

Table 3: Resulting time on optimised solution with the initial rider configuration ABC

Best race time (s)	Mean time (s)	Standard deviation	Rider A final energy (J)	Rider B final energy (J)	Rider C final energy (J)
204.52	205.403	0.482	35	96	1948

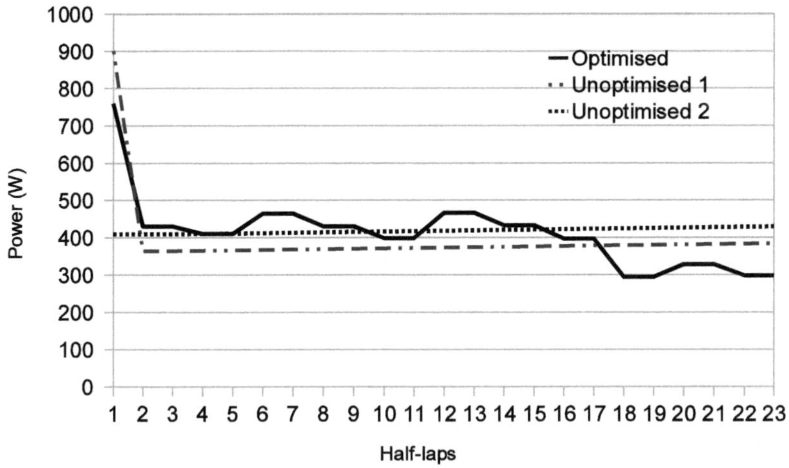

Fig. 3: Comparison of the optimised solution's power profile to that of the two unoptimised solutions, all under the same strategy

Table 4: Resulting time on optimised solution with all configurations

Configuration	Best race time (s)	Mean time (s)	Standard deviation
ABC	204.52	205.403	0.482
ACB	204.82	205.608	0.486
BAC	204.82	205.507	0.394
BCA	203.32	204.006	0.383
CAB	205.22	205.947	0.452
CBA	204.32	204.830	0.327

investigates the optimised solution with the selected strategy, based on the other permutations of the initial rider configuration. This is again performed with 100 repetitions and 2000 evaluations per repetition.

We can see that we are able to return a better time by altering the initial configuration of the riders. Here, BCA gives a better time of 203.32 with the power profile [760, 404, 467, 432, 395, 467, 435, 393, 471, 411, 385, 126]. This is interesting, as it may be intuitive to predict that placing the rider with the most energy in the initial high power phase would be optimal. It is also interesting to note the low standard deviation, which shows a relatively consistent race time is returned.

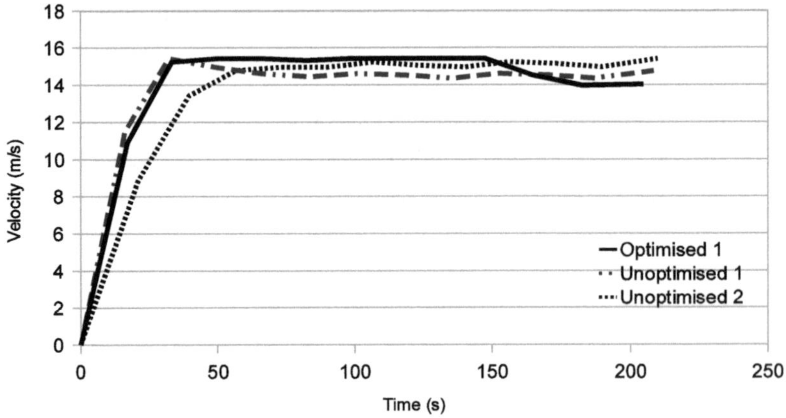

Fig. 4: Comparison of the optimised solution's velocity profile to that of the two unoptimised solutions, all under the same strategy

Table 5: Optimising transition strategy and power profile for all rider configurations

Algorithm	Configuration	Best race time (s)	Mean time (s)	Standard deviation
RLS	ABC	202.08	202.881	0.864
RLS	ACB	202.08	202.861	0.628
RLS	BAC	202.20	202.841	0.507
RLS	BCA	201.96	202.945	0.712
RLS	CAB	202.20	203.180	0.829
RLS	CBA	202.10	203.198	0.837
Simple EA	ABC	201.98	203.084	0.656
Simple EA	ACB	202.20	203.101	0.610
Simple EA	BAC	201.90	203.203	0.683
Simple EA	BCA	202.02	202.688	0.385
Simple EA	CAB	202.20	203.049	0.542
Simple EA	CBA	202.22	203.087	0.614

5.3 Strategy Optimisation

After demonstrating the decreased race time gathered from the power optimisation, we now demonstrate the effectiveness of the combined transition strategy and power optimisation. Running RLS and Simple EA over the six permutations of initial rider configuration returned the results in Table 5. These results were obtained by performing 100 repetitions each with 2000 CMA-ES evaluations. The Simple EA was run for 100 EA evaluations.

We can see that an evolutionary approach to the transition strategy and power profile is able to further speed the race, allowing consistent times of 202 s. The BAC-based strategy evolved by the Simple EA with the transition strategy [1, 2, 3,

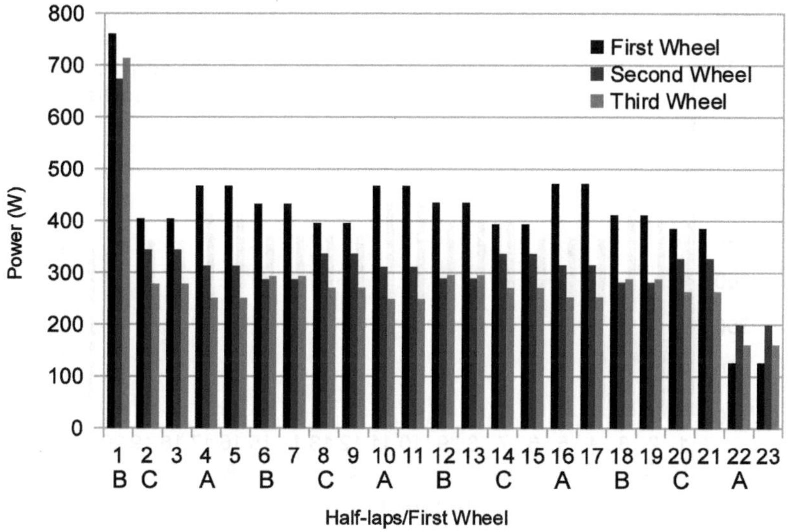

Fig. 5: The power profile of the best power-optimised solution, BCA

3, 2, 3, 1, 2, 3, 2, 1] and the power profile [767, 464, 397, 430, 470, 400, 400, 449, 387, 373, 105] is able to reach the best time of 201.90 s, 1.42 s faster than the solely power-optimised BCA. Figures 5 and 6 demonstrate the power profile of all three riders for both the power-optimised BCA and the strategy-optimised BAC. Figure 7 compares the two velocity profiles; it is interesting to note how little the two profiles vary for a 1.42 s difference in time.

5.4 Statistical Significance of Results

It was observed in Sect. 5.3 that both the Simple EA and the RLS algorithms were able to produce a superior time than the best power-optimised experiment. In this section, we analyse the results of the combined power and strategy optimization techniques in order to determine which is superior.

It must be noted that the results of the tests can not be assumed to be normally distributed, and that it is unknown before analysis which algorithm provides the better results. In order to determine which is the better algorithm, an unpaired and non-parametric statistical hypothesis test is required. The results of the RLS and Simple EA thus are compared using the Mann-Whitney-Wilcoxon test with a two-sided *P*. This method ranks the times of the experiments, and these ranks form a normal distribution. Using these ranks, we can say whether a difference exists between the two sets of results. The null hypothesis states that there exists no difference between the RLS and Simple EA results, and the accompanying alternate hypothesis simply states that either the RLS or Simple EA is superior.

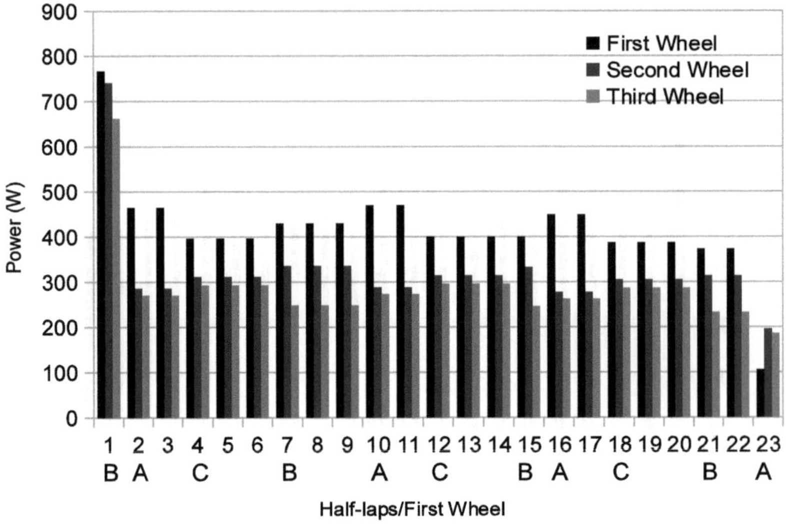

Fig. 6: The power profile of the best strategy-optimised solution, BAC under Simple EA

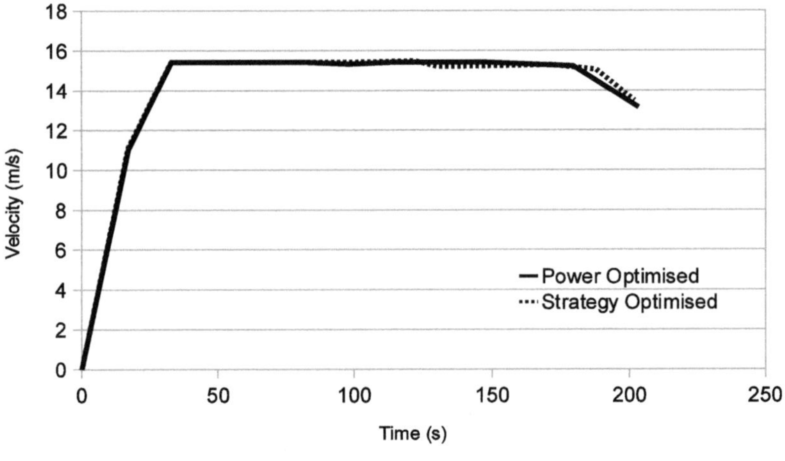

Fig. 7: Comparison of the velocity profiles of the power-optimised BCA (203.32 s) and the Simple EA strategy-optimised BAC (201.90 s)

Note that lower values are superior to higher values in this analysis, and hence larger ranks will correspond to superior values. This means that the algorithm returning faster values will have a greater area under the curve.

Table 6 compares RLS to Simple EA with the results combined, giving a sample size of 600 per algorithm. Rider permutations are not considered in this table, and the mean results are obtained from 600 experiments per algorithm. U_{RLS} provides

Table 6: Overall Mann–Whitney–Wilcoxon comparison on the RLS and Simple EA experiments

Mean_{RLS}	$\text{Mean}_{Simple\ EA}$	U_{RLS}	AUC_{RLS}	p	Significant
203.628	203.782	218066.5	0.606	<0.0001	+

Table 7: Mann–Whitney–Wilcoxon comparison on the individual RLS and Simple EA experiments

Perm.	Mean_{RLS}	$\text{Mean}_{Simple\ EA}$	U_{RLS}	AUC_{RLS}	p	Significant
ABC	202.881	203.084	6537.5	0.654	0.0002	+
ACB	202.861	203.101	6419.5	0.642	0.0005	+
BAC	202.841	203.203	6698.5	0.670	<0.0001	+
BCA	202.945	202.688	4207	0.421	0.0524	−
CAB	203.180	203.049	4997.5	0.500	1	−
CBA	203.198	203.087	4981	0.498	0.9601	−

the Mann-Whitney U value for RLS, and AUC_{RLS} provides the area under the curve for the RLS. A proportion over 0.5 shows that the resulting times of RLS are ranked lower than those of the Simple EA, meaning they are to some degree faster than those of the Simple EA. p provides the significance value of the results; the results are considered significant and the null hypothesis rejected if $p < 0.05$. Significant analyses are denoted as such with a $+$ in the Significant column.

As can be seen, the results of the Mann-Whitney-Wilcoxon test are statistically significant, showing that the RLS and Simple EA results are different. In addition, observing the proportion of the area under the curve, RLS has returned a superior set of times.

In order to analyse the component experiments, the Mann-Whitney-Wilcoxon test is applied to each of the permutations. Table 7 provides the results of the comparison between each of the RLS and Simple EA experiments under a sample size of 100 per test. The rider permutations and the mean results for the two algorithms are as per Table 5. As can be seen, three of the permutations show that RLS is significantly different than the Simple EA, and the proportion of the area under the curve shows that RLS has returned a superior set of times. The other three tests are not significant, with two providing a p-value of over 0.95.

This analysis has shown that the results from RLS range from equivalent to superior, showing that the RLS algorithm is the current optimal choice.

6 Conclusions and Future Work

In this paper, we addressed the problem of evolving team strategies for track cycling. We introduced a model of a team consisting of cyclists with different properties, such as weight, maximum power output, and available amount of energy.

This allowed us to evaluate strategies without having athletes explicitly evaluating the strategies on track.

The experiments revealed several interesting insights. It is important to note that it is undesirable to manually generate the power profile for a race, and that evolutionary algorithms may be used to reduce a race time by optimising the power profile. However, all permutations of initial rider configuration should be investigated because it may not be obvious which configuration of riders is optimal for a particular race. Finally, an algorithm that optimises both transition strategy and power profile is able to obtain a superior race time. The RLS and Simple EA algorithms were able to return a considerably faster time than either the unoptimised or transition-strategy optimised solutions for the domain.

In the future, we will continue our research in the following areas:

- Validation and improvement of the obtained solutions through cooperation with professional cyclists.
- Investigation into the sensitivity of the obtained solutions to parameter variations. As there may be variations or errors in parameter measurement, it would be of interest to produce a stochastic model that gives the confidence range of a given solution. An additional objective would be to then reduce the standard deviation of the stochastic model's distribution of results, thus creating a multi-objective optimisation, as we want a low race time but also high confidence.

Acknowledgements The authors would like to thank Dr David Martin from the Australian Institute of Sport cycling program and Dr Tammie Ebert from Cycling Australia for their valuable support.

References

1. Bunte, S., Kliewer, N.: An overview on vehicle scheduling models. Publ. Transport **1**, 299–317 (2009)
2. de Koning, J.J., Bobbert, M.F., Foster, C.: Determination of optimal pacing strategy in track cycling with an energy flow model. J. Sci. Med. Sport **2**(3), 266–277 (1999)
3. Hansen, N.: The CMA evolution strategy: a comparing review. In: Lozano, J., Larranaga, P., Inza, I., Bengoetxea, E. (eds.) Towards a New Evolutionary Computation. Advances in Estimation of Distribution Algorithms, pp. 75–102. Springer, Heidelberg (2006)
4. Herrera, F., Lozano, M., Verdegay, J.L.: Tackling real-coded genetic algorithms: Operators and tools for behavioural analysis. Artif. Intell. Rev. **12**(4), 265–319 (1998)
5. Martin, J.C., Gardner, A.S., Barras, M., Martin, D.T.: Modeling sprint cycling using field-derived parameters and forward integration. Med. Sci. Sports Exerc. **38**(3), 592–597 (2006)
6. Olds, T., Norton, K., Lowe, E., Olive, S., Reay, F., Ly, S.: Modeling road-cycling performance. J. Appl. Physiol. **78**(4), 1596 (1995)
7. Shabtay, D., Steiner, G.: A survey of scheduling with controllable processing times. Discrete Appl. Math. **155**(13), 1643–1666 (2007)
8. Talbi, E.G.: Metaheuristics: From Design to Implementation. Wiley, New York (2009)

A Dual Mutation Operator to Solve the Multi-objective Production Planning of Perishable Goods

Pedro Amorim, Carlos Henggeler Antunes, and Bernardo Almada-Lobo

Abstract This work proposes a dual mutation operator that uses dual information to solve the multi-objective lot-sizing and scheduling problem arising in perishable goods production planning. This operator takes advantage of the hybrid structure of a multi-objective evolutionary algorithm, which combines the well-known Non-Dominated Sorting Genetic Algorithm and a mixed-integer linear programming solver. Chromosomes are coded as strings of integer values fixing the production sequence. For each feasible individual, the reduced costs of the relaxed binary variables coming from the coding scheme are used to guide the mutation process. To assess the performance of this operator, a set of randomly generated instances based on a methodology reported in the literature is tested and evaluated with multi-objective performance metrics. The results indicate that the operator is able to achieve consistently better solutions in terms of an approximation to the Pareto solutions compared to the same solution procedure without the dual mutation.

1 Introduction

Increasing attention has been paid to tackle product perishability in supply chain planning and methods able to deal with models which explicitly consider it are needed [1, 4]. The effects of final product perishability in production planning and scheduling impact various performance indicators, such as the inventory levels, the number of setups performed or the customer satisfaction.

Pedro Amorim · Bernardo Almada-Lobo
INESC TEC, Faculdade de Engenharia, Universidade do Porto, Porto, Portugal
e-mail: amorim.pedro, almada.lobo@fe.up.pt

Carlos Henggeler Antunes
Department of Electrical Engineering and Computers and INESC Coimbra, University of Coimbra, Coimbra, Portugal
e-mail: ch@deec.uc.pt

L. Di Gaspero et al. (eds.), *Advances in Metaheuristics*, Operations Research/ Computer Science Interfaces Series 53, DOI 10.1007/978-1-4614-6322-1_5,

Simultaneous lot-sizing and scheduling is mandatory to be executed whenever the underlying production process has sequence-dependent setups [2]. This is the case in many production systems of food consumer goods, such as the yoghurt production [24]. When considering lot-sizing and scheduling as separate problems, several works recognize the importance of a multi-objective approach to tackle conflicting objectives. This is not the case in the works dealing with these problems in an integrated manner. However, in case freshness needs to be accounted for when dealing with perishable products, a multi-objective framework is an advantageous modelling technique with significant benefits for supporting the decision maker [3]. In fact, besides avoiding the assignment of fictitious costs related to freshness loss, which are less realistic, the multi-objective approach offers the decision maker a tool to trade-off explicitly total costs against product freshness.

When solving a multi-objective lot-sizing and scheduling problem, it is important to take into account three features of the problem structure. First and foremost, single-objective lot-sizing and scheduling has been proven to be NP-complete, even for finding a feasible solution [9]. The multi-objective version will be at least as hard as the single-objective version. Second, in a multi-objective problem two important characteristics of the Pareto-front should be satisfied: a well-spread set of solutions and a front that is as near as possible to the optimal. Third, we should be aware that lot-sizing and scheduling problems have mainly two types of variables: continuous variables related to the size of the lots to be produced and binary variables related to the scheduling of those lots. In this paper we show how a multi-objective evolutionary algorithm (MOEA) [6] coupled with a mixed-integer linear programming (MIP) solver can be used to solve this problem by exploring these three features. Additionally, we focus on the potentialities of this hybridization by proposing a dual mutation operator that takes advantage of the information extracted from the MIP solver in order to guide the search [14]. This operator acts on feasible individuals assigning tailored mutation probabilities to each gene based on the reduced costs of the relaxed binary variables coded in the chromosomes. Thus, we aim at exploring each individual by guiding its path through the solution space.

The remainder of the paper is organized as follows. The next section is devoted to the literature review on similar problems and solution methods. Section 3 outlines the problem definition and its mathematical formulation. In Sect. 4, a detailed description of the MOEA and the dual mutation operator is given. In Sect. 5, the results of computational experiments are provided. Finally, in Sect. 6 conclusions are drawn and future research directions are pointed out.

2 Literature Review

The present work intersects two research areas that have been identified as being still in their infancy: the production planning and scheduling of perishable goods and the development of hybrid multi-objective algorithms that combine metaheuristics and exact methods.

In the recent paper [20] and in the review [4], the key papers in the field of production planning and scheduling of perishable goods are enumerated and analysed. Marinelli et al. [25] studies the production planning in the yoghurt industry. The problem is formulated as a special case of the capacitated lot-sizing and scheduling problem with parallel machines, which includes shared buffers containing the same recipe. The objective function is typical for production planning problems and aims to minimize inventory, production and setup costs. In order to solve this hard combinatorial problem, a two-step heuristic that decomposes the problem into a lot-sizing and a scheduling problem is developed. Still regarding the planning and scheduling of the packaging stage in the yoghurt industry, [24] presents three mixed-integer linear models that incorporate shelf-life issues. These models use the concept of families/blocks to simplify the scheduling decisions. All models are solved recurring to a state-of-art MIP solver. Pahl and Voß [28] and Pahl et al. [29] extend well known discrete lot-sizing and scheduling models, such as the general lot-sizing and scheduling problem by including deterioration and perishability constraints. The impact of minimum lot-sizes on the number of spoiled products is well documented. Amorim et al. [3] tackles the production planning of perishable products using a multi-objective framework, trading-off costs and freshness. Two models were proposed depending on the production strategy and a number of real-life dairy instances are solved. The solution method used consists of a hybrid multi-objective evolutionary algorithm (MOEA). The works [19, 22, 21, 20] propose general MIP formulations for food processing industries ranging from yoghurt to ice-cream plants. These models have multi-stage considerations and several industry related constraints, such as natural setup sequences, sequence-dependent changeover times and costs, and the possibility of production overtimes. The latter papers focus also on the development of valid inequalities that improve the performance of solvers in finding solutions for real-life instances. More focused on scheduling, the work [11] presents various heuristics to solve the problem arising at the packing lines in dairy industries.

Concerning hybrid multi-objective algorithms, the work [8] reviews procedures that combine metaheuristics and exact methods. The authors acknowledge that despite the fact that in single-objective optimization the amount of cross-fertilization work has been rising interestingly, "it is astonishing to observe that this way to proceed is marginal in the multi-objective context". Nevertheless, among the few papers taking advantage of this hybridization, different motivations can be found underlying the innovative search procedures. In [10] the idea is to reduce the decision search space of the multi-objective knapsack problem with an exact procedure and afterwards running an adapted tabu search on a potentially more efficient region. The cuts applied ensure that no optimal solutions are eliminated and, therefore, in this solution method the exact methods help in enhancing the metaheuristic performance. Contrarily, [31] proposes a solution procedure for the multi-objective assignment problem in which a population based heuristic using path-relinking is used to accelerate an implicit enumeration scheme. Concerning the multi-objective knapsack problem, [13] combines heuristic search strategies that are used to detect potentially good regions in the decision search space, which are then explored by an exact method. Following the same reasoning of using exact methods to solve sub-problems to optimality, [17] addresses the bi-objective covering tour problem using a MOEA combined with a branch-and-cut algorithm.

The literature reviews helps to make clearer the contribution of this work. First, it consolidates the importance of more work in the production planning and scheduling of perishables goods. Second, it proposes a dual mutation operator that takes advantages of the hybridization between the MOEA and the exact solver. This field of research is incipient and, therefore, needs actual proofs of the improvement in the results that can be derived from investing in hybrid methods capable of making the most of metaheuristics and exact methods.

3 Problem Definition

This paper considers the lot-sizing and scheduling problem of perishable products. The main characteristic of perishable products is their relative short shelf-life. Our focus is on the semi-continuous production process of perishable consumer goods. Their production environment is characterized by a continuous production stage, followed by a packaging stage, which is known in the literature as "make-and-pack" production.

Lot-sizing and scheduling constitutes one of the major challenges in this type of production environment, in which a large number of products is produced from a few initial product recipes. Thus, in these settings, there is a wide variety of products that differ in small features, such as the size of the container, the language on the label, or the flavour. In this work we are only concerned with the packaging stage where the divergent product structure (i.e., a low number of raw materials leading to a high number of final products) shows its impact by imposing to the planner the need to deal with a great amount of stock keeping units to size and schedule.

In this kind of consumer goods industries, recipes are usually subject to empirical sequencing rules based on colours, flavours or temperatures that push the planners towards the concept of product family. This planning feature is much related to the block planning concept, also called production wheel policy, which will be partially used in our formulation [15]. Hence, if we consider, for example, the recipe of high fat yoghurt that has a setup time dependent on the recipe that was produced before, the product sequencing within this recipe is pre-defined and always starts with the plain yoghurt and ends with the chocolate flavour.

Due to the great diversification that has been imposed to these goods by the market and their rather short-life, many different products must be produced on a daily basis, which increases the complexity in making scheduling decisions. Moreover, in order to account for perishability, it is important to simultaneously know when and how much to produce in a weekly rolling-planning horizon.

3.1 Mathematical Model

All product variants $k = 1, \ldots, K$ based on the same recipe form a block; therefore, a product can be assigned to one block only. Blocks $j = 1, \ldots, J$ are to be scheduled on $l = 1, \ldots, L$ parallel production lines over a finite planning horizon consisting of macro-periods $d, h = 1, \ldots, D$ with a given length (in this case a day). The scheduling takes into account the sequence dependent setup time and cost between blocks. The sequence of products for a given block is set a priori due to natural constraints in this kind of industries. Hence, when changing the production between two products that are variants of the same block/recipe solely a minor setup is needed, which is not dependent on the sequence but on the product to be produced.

A macro-period is divided into a fixed number of non-overlapping micro-periods with variable length. Since the production lines can be independently scheduled, this division is done for each line separately. S_{ld} denotes the set of micro-periods s belonging to macro-period d and production line l. All micro-periods are put in order $s = 1, \ldots, S^l$, where S^l represents the total number of micro-periods on line l implicitly defined by S_{ld} ($S^l = \sum_d |S_{ld}|$).

The length of a micro-period is a decision variable that depends on the production amount of several products of the same block. A sequence of consecutive micro-periods, where the same block is produced on the same line, defines the lot of a given block. Therefore, a lot may aggregate several products from a given block and may continue over several micro and macro-periods. Moreover, a lot is independent of the discrete time structure of the macro-periods. The number of micro-periods of each day defines the upper bound on the number of blocks to be produced daily on each line. As a consequence of the parameter S_{ld}, a lot may contain idle micro-periods with null production. In case, the same block is produced after an idle micro-period, on the same line, the setup state is conserved. Note that this lot-sizing and scheduling structure is based on the general lot-sizing and scheduling structure for parallel lines [26].

Consider the following indices, parameters and decision variables.

Indices

l	Parallel production lines
i, j	Blocks
k	Products
d, h	Macro-periods
s	Micro-periods

Parameters

K_j	Set of products belonging to block j
$\|K_j\|$	Number of products belonging to block j
S_{ld}	Set of micro-periods s within macro-period d and line l
$[d_{kd}]$	Number of non-zero occurrences in the demand matrix
Cap_{ld}	Capacity (time) of production line l available in macro-period d

a_{lk}	Capacity consumption (time) needed to produce one unit of product k on line l
c_{lk}	Production costs of product k (per unit) on line l
u_k	Shelf-life duration of product k after completion of its production (time)
m_{lj}	Minimum lot-size (units) of block j if produced on line l
$sc^b_{lij}(st^b_{lij})$	Sequence dependent setup cost (time) of a changeover from block i to block j on line l
$sc^p_{lk}(st^p_{lk})$	Sequence independent setup cost (time) of a changeover to product k on line l
d_{kd}	Demand for product k in macro-period d (units)
y_{lj0}	Equals 1, if line l is set up for block j at the beginning of the planning horizon (0 otherwise)

Decision Variables

w_{khd}	Fraction of demand of product k produced in macro-period h for meeting demand in macro-period d, $h \leq d \leq h + u_k$
q_{lks}	Quantity of product k produced in micro-period s on line l
p_{lks}	Setup state: $p_{lks} = 1$, if line l is set up for product k in micro-period s (0 otherwise)
y_{ljs}	Setup state: $y_{ljs} = 1$, if line l is set up for block j in micro-period s (0 otherwise)
z_{lijs}	Takes on 1, if a changeover from block i to block j takes place on line l at the beginning of micro-period s (0 otherwise)

We highlight that variable w_{khd} is only instantiated for $h \leq d \leq h + u_k$ to ensure that demand fulfilled in period d is produced beforehand with units that have not perished yet. It is assumed that the inventory at the beginning of the planning horizon is null, and that no stock is kept at the end of it. Nevertheless, intermediate units might be held when variable w_{khd} is positive for $d > h$.

The lot-sizing and scheduling of perishable goods (MO-LSP) may be formulated as a multi-objective MIP model as follows:

MO-LSP

$$\min f^1 = \sum_{l,i,j,s} sc^b_{lij} z_{lijs} + \sum_{l,k,s} (sc^p_{lk} p_{lks} + c_{lk} q_{lks}) \tag{1}$$

$$\max f^2 = \frac{1}{[d_{kd}]} \sum_{k,h,d} \frac{h + u_k - d}{u_k} w_{khd} \tag{2}$$

subject to:

$$\sum_h w_{khd} = 1 \quad \forall k,d : d_{kd} > 0 \tag{3}$$

$$\sum_h w_{khd} = 0 \quad \forall k,d : d_{kd} = 0 \tag{4}$$

$$\sum_{l,s \in S_{lh}} q_{lks} \geq \sum_{d} w_{khd} d_{kd} \quad \forall k,h \tag{5}$$

$$\sum_{k \in K_j} p_{lks} \leq y_{ljs} |K_j| \quad \forall l,j,s \tag{6}$$

$$q_{lks} \leq \frac{Cap_{ld}}{a_{lk}} p_{lks} \quad \forall l,k,d,s \in S_{ld} \tag{7}$$

$$\sum_{i,j,s \in S_{ld}} st^b_{lij} z_{lijs} + \sum_{k,s \in S_{ld}} (st^p_{lk} p_{lks} + a_{lk} q_{lks}) \leq Cap_{ld} \quad \forall l,d \tag{8}$$

$$\sum_{j} y_{ljs} = 1 \quad \forall l,s \tag{9}$$

$$\sum_{k \in K_j} q_{lks} \geq m_{lj}(y_{ljs} - y_{lj,s-1}) \quad \forall l,j,s \tag{10}$$

$$z_{lijs} \geq y_{li,s-1} + y_{ljs} - 1 \quad \forall l,i,j,s \tag{11}$$

$$(w_{khd}, q_{lks}, z_{lijs}) \geq 0, \ (p_{lks}, y_{ljs}) \in \{0,1\} \tag{12}$$

The first objective is related to the minimization of production costs and it has an intrinsic combinatorial nature, while the second objective aims at maximizing the freshness of the products delivered and has a linear structure. The freshness objective takes advantage of the simple plant location reformulation used in the production variables that determine explicitly when a product is produced and delivered. Taking into account the shelf-life information for each product, the goal is to maximize the average percentage remaining shelf-life of all requests. In this expression, different remaining shelf-lives between products are all normalized between 0 and 1. With Eqs. (3) and (4) the demand for all products is fulfilled and with constraints (5) the demand requirements are translated into production quantities. Constraints (6) and (7) ensure that in order to produce a certain product, the necessary setup for it and for the corresponding recipe has to be performed, respectively. Moreover, at each moment only products from the same recipe may be produced Eq. (9) and they are subject to a minimum batch size Eq. (10). Constraints (8) limit the use of the capacity in each macro-period and, finally, constraints (11) are responsible for tracking the changeover between blocks. For more details about the mathematical model the readers are referred to [3].

4 Multi-objective Hybrid Genetic Algorithm

To solve this problem we use an algorithm similar to the one presented in [3], which consists of a hybridization of a MOEA and an MIP solver. This was the first time that an MO-LSP model had been proposed. The underlying MOEA is the Non-Dominated Sorting Genetic Algorithm (NSGA-II), which has proven to perform rather well in comparison to other MOEAs [18]. The readers are referred to [7] for the details of NSGA-II. In this work, we further exploit the potentialities of such hybridization by developing a new dual mutation operator that acts on feasible individuals. It is based on the individual values of the reduced costs from the variables related to the chromosome representation. Generally, standard mutation operators have a myopic view about the differences on the information coded across the population. With this operator this limitation is overcome by giving tailored mutation probabilities for each gene. In Fig. 1, an outline of the proposed hybrid algorithm is shown with the dual mutation operator used to solve the MO-LSP (the meaning of f^3 is clarified in Sect. 4.5).

Pedro Amorim, Carlos Henggeler Antunes, Bernardo Almada-Lobo

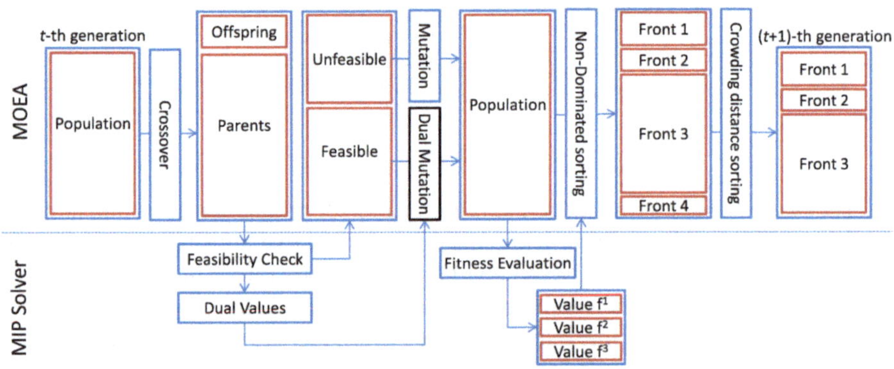

Fig. 1: Hybrid genetic multi-objective algorithm with dual mutation

There are five important components that are needed to completely describe any genetic algorithm (GA) [12]: the representation format used by the GA; the genetic operators; the fitness evaluation function; the various parameters, such as the population size, the number of generations, the probability of applying the operators, etc.; and a method for generating the initial population. As far as our approach is concerned, it is also important discuss about infeasibility/constraint handling issues to increase the flexibility of the solution space search. The parameters will be described in Sect. 5.3.

4.1 Representation of an Individual

The usual representation of a solution in a GA for a lot-sizing and scheduling problem is a string of paired values for each scheduling period in the planning horizon [30]. The first value indicates the type of product and the second value the number of units of that product type to be produced, i.e., the lot-size. Due to our hybrid approach, an individual is composed of a string of single values. The value of each gene represents the block to be scheduled in each micro-period of a given machine. Therefore, there are $\sum_l S^l$ genes in each chromosome. The lines are put in numerical order so that at every position it is possible to return exactly the micro-period and the machine accessed. This yields an incomplete representation of the individual, as it only determines the value of the y_{lks} variables. The values of the remaining variables are obtained in a subsequent step by means of the MIP solver through an aggregated scalar objective function.

4.2 Genetic Operators

Four types of genetic operators are implemented as follows.

4.2.1 Crossover and Standard Mutation

Since it is usual in these industries to have dedicated lines, sets of blocks/products may only be produced on certain machines. Hence, one needs to ensure that the changes performed by the crossover operator respect this condition. Here, the one-point crossover is used and, consequently, when combining the genetic material of the two parents, we are sure that the exchanges occur for the same lines. Therefore, we cut the chromosomes by a given micro-period, which belongs to a line and a macro-period, and obtain the child population based on these cut points.

Regarding the standard mutation operator, it is similar to the bitwise mutation but, in this case, a gene is randomly selected, which represents a block produced in a given micro-period, and its value is changed into a value corresponding to a block that is allowed to be produced on that line. This standard mutation operator is only applied on infeasible individuals after the execution of a feasibility check.

4.2.2 Dual Mutation

This mutation operator only acts on feasible individuals. As input it uses the reduced costs, $\bar{y}_{ljs}$, of the y_{ljs} variables. Hence, if the MIP associated with a certain individual yields a feasible solution, the algorithm fixes all integer variables of the solution to the values obtained from the solution representation and the MIP solver. After, the corresponding linear programming (LP) model is solved in order to obtain $\bar{y}_{ljs}$. Note that $\bar{y}_{ljs}$ can be obtained by solving the following equations:

$$\bar{y}_{ljs} = co_{ljs} - \sum_{ct \in M} ro_{ct,ljs} sp_{ct} \quad \forall l, j, s$$

where co_{ljs} represents the implicit coefficients of y_{ljs} in the aggregated objective function, M is the set of all constraints ct in which y_{ljs} appears, sp_{ct} is the shadow price associated with constraint ct and $ro_{ct,ljs}$ is the coefficient of y_{ljs} in constraint ct. Thus, if we think of co_{ljs} as marginal revenues, the reduced cost $\bar{y}_{ljs}$ represents the net marginal revenue associated with variables y_{ljs} in the LP relaxation.

$\bar{y}_{ljs}$ can take three possible types of values: (i) positive, if the entry of such variable to the base would increase the objective function value; (ii) negative, if the entry of such variable would decrease the objective function value; or (iii) 0, if the variable is already a basic variable of the solution (as fixed through the chromosome representation) or the entry of such variable would not change the objective value. Recall that these straightforward conclusions are valid for the LP relaxation. The objective functions in the related MIP are not influenced in such a direct way because y_{ljs} are binary variables.

Let us now focus on the meaning of these values regarding our specific problem and define α, β, γ as the weighting coefficients associated with each of $\bar{y}_{ljs}$ possible value (positive, negative and 0, respectively). The probability of mutation to the respective variable is proportional to these waits. Since both objective functions are coded in the algorithm as minimizations, in case $\bar{y}_{ljs} > 0$ we want to avoid that the associated y_{ljs} variables take the value of 1 in individuals of a future generation. Therefore, the value of α should remain close to zero in order to prevent such situation. Looking deeper at the case where $\bar{y}_{ljs} < 0$, we should note that concerning the cost minimization function, an increase of one unit in the lower bound of any y_{ljs} would almost never result in a negative reduced cost as it would probably augment or maintain this objective function value through the execution of more setups. We highlight that inventory costs are not incorporated in the objective function f^1. Therefore, the usual trade-off between holding and setup costs does not play a role here. Thus, negative values of $\bar{y}_{ljs}$ will be linked with the objective function related to the maximization of freshness. Therefore, β is related to the contribution of the dual mutation operator in attaining better values of freshness. Finally, when $\bar{y}_{ljs} = 0$, the impact of having the corresponding y_{ljs} as a basic variable is not clear, but it might turn out to be beneficial in a future generation. Consequently, γ should have a significant value so that diversification is accounted for.

The transformation that each feasible individual undertakes from the dual mutation operator is depicted in Fig. 2. The weights in this example are set as follows: $\alpha = 0$, $\beta = 0.5$ and $\gamma = 1$. The probabilities are calculated by dividing the weight of a given block over the sum of all weights.

To understand the possible path of a gene let us consider, for example, gene 2 (that corresponds to the second micro-period of the first macro-period) that is setting up the machine for producing block **3** close to the beginning of the planning horizon. First, the reduced costs for each block that can be set up are calculated. The reduced cost related to block 4 is positive and, therefore, the final probability assigned to it is 0. With regard to block 5, its negative reduced cost would probably induce an increase of the freshness value. The probability of 0.14 of this block is computed from the following expression: $0.5/(1+1+1+0.5)$. In the end, this gene mutates to either block 1, 2, or 3 with equal probability of 0.29 and to block 5 with probability of 0.14 (in Fig. 2 the result is a mutation to block **5**).

Dual mutation operator to solve the production planning of perishable goods

Old Individual	j	2	3	5	1	1	4	4	4	2	5
y_{ljs}	1	-20	0	10	0	0	-5	10	0	20	-10
	2	0	0	0	20	0	0	-5	0	0	-10
	3	0	0	0	-5	0	-20	0	10	10	0
	4	-10	10	1	20	-5	0	0	0	10	-10
	5	5	-20	0	5	10	20	5	5	20	0
Weights	1	0.5	1	0	1	1	0.5	0	1	0	0.5
	2	1	1	1	0	1	1	0.5	1	1	0.5
	3	1	1	1	0.5	1	0.5	1	0	0	1
	4	0.5	0	0	0	0.5	1	1	1	0	0.5
	5	0	0.5	1	0	0	0	0	0	0	1
Probability	1	0.17	0.29	0.00	**0.67**	**0.29**	0.17	0.00	**0.33**	0.00	0.14
	2	**0.33**	0.29	0.33	0.00	0.29	0.33	0.20	0.33	**1.00**	**0.14**
	3	0.33	0.29	0.33	0.33	0.29	0.17	0.40	0.00	0.00	0.29
	4	0.17	0.00	0.00	0.00	0.14	**0.33**	**0.40**	0.33	0.00	0.14
	5	0.00	**0.14**	**0.33**	0.00	0.00	0.00	0.00	0.00	0.00	0.29
New Individual		2	5	5	1	1	4	4	1	2	2

Fig. 2: Schematic steps of the dual mutation operator. The probabilities randomly selected to perform the corresponding family mutation are in *bold*

4.2.3 Selection/Reproduction

After the offspring population is created, the whole population is sorted according to the non-dominance criterion. In the end of this process every individual belongs to a front. The new parent population is formed by adding solutions from the best fronts until the number of individuals selected reaches a pre-defined parameter of the size of the population. Thereafter, the solutions of the last front (the worst front accepted from the non-dominance ranking) are sorted according to the crowding operator. This operator conducts an internal ranking of the last front based on the distance of each individual to its closest neighbours. Individuals with the *farthest* neighbours are then selected until the rest of the new population is filled. Clearly, this operator favours a well-dispersed set of solutions.

4.3 Fitness of an Individual

The hybridization of the NSGA-II with the MIP solver is also done in this step. Each individual representing a sequence of blocks to be scheduled on each line is fed into the MIP model by fixing the values of the y_{lks} variables and, consequently, a significant number of p_{lks} as well. Notice that y_{lks} and p_{lks} are the only binary variables of our mathematical model. The sub-problem is much easier for the MIP solver since optimal lot-sizes can be computed based on the fixed block sequence of the individuals while minimizing a weighted linear combination of both objectives. Each time an individual is fed into the solver, the two objectives are randomly combined with a weighted sum leading to a scalar objective function. This random weighting aims at providing a good dispersion of results throughout the Pareto-front, since with fixed weights the solution tends to converge quickly to a certain region of the solution space and be entrapped in it. The objective function values are then used to rank individuals according to non-dominance and to compute the Pareto-fronts. If the individual turns out to be infeasible, then a feasibility recovery procedure is applied, as explained in Sect. 4.5.

4.4 Initializing a Population

A population is initialized by randomly generating values for each chromosome. Nevertheless, special attention is paid to be sure that the possible values are in compliance with the set of blocks/products that can be produced on a line. This way, we ensure from the beginning that individuals respect the allowed sequences on every machine. Moreover, diversity within the population is also guaranteed.

4.5 Infeasible Individuals

Infeasible individuals are neither automatically disqualified from the population, nor are they repaired to create feasible individuals. Instead, the amount of infeasibility is defined as an additional objective that must be minimized. Justification for treating the infeasibility in this way is provided in [27]. This approach has been suggested by [5], as a way of avoiding the definition of a more complex, but constantly feasible representation, specifically designed operators, or a repair method. Furthermore, since the optimal solutions may lie close to the infeasible region, this approach allows these frontier regions to be well explored.

Therefore, when trying to calculate the fitness of an individual regarding both objectives, in case the individual is infeasible, a new objective is set as $\min f^3 = \sum_{ct} v_{ct}$, where v_{ct} is the amount of violation of constraint ct (cf. Fig. 1). In order to measure this violation, slack variables are introduced and their values are to be minimized expecting that the individual will turn feasible in a future generation by means of crossover and mutation.

5 Computational Experiments

In this section, the computational performance of the algorithm with and without the dual mutation operator is assessed. To test the proposed method, a C++ implementation with the MIP solver CPLEX 12.1 is run.

5.1 Data Generation

A total of 27 instances were systematically generated, following a methodology very similar to [16]; therefore, L was set to 1. For all products $a_{lk} = 1$ and the machine is set up for block 1 in the beginning of the planning horizon. The number of blocks J is 5, 10 and 15, and each block has only one product k. The number of macro-periods D is 5, 10 and 20. The number of micro-periods within a macro-period S_{ld} is set to the value of J allowing all products to be produced in each macro-period with minimum lot-sizes m_{lj} of 1 unit. A demand matrix with 15 products (rows) and 20 macro-periods (columns) is randomly generated, where each element d_{kd} is in the interval [40, 60] of the uniform distribution. For the setup times between blocks (st^b_{lij}) the interval [2,10] is used (except for the case where $i = j$, where the setup is 0). For the setup of products it was considered that both the respective time and cost as well as production costs are null. Shelf-lives u_k are generated for all 15 products for each possible planning period length choosing randomly from the interval [1, D]. Hence, for an instance with J blocks and D macro-periods we used the data given in the first J rows and the first D columns of the demand matrix, the first J rows and columns of the setup time, and the first J entries of the shelf-life vector built for D macro-periods. Then, as in [16], the concept of common random number is present in our experiments. The setup cost sc^b_{lij} for a changeover from block i to block j is computed as:

$$sc^b_{lij} = f_{sc} st^b_{lij} \quad \forall l,i,j.$$

Although in [16] the parameter f_{sc} is varied between 50 and 500, in our case, due to the structure of our algorithm, the impact of this parameter in the efficiency is null. The relation between setup times and costs does not influence the performance of our algorithm at all, because the sequencing decision is explicitly defined from the individual representation. Hence, this parameter is fixed to 50 for every instance. The capacity per macro-period Cap_{ld} is determined according to:

$$Cap_{ld} = \frac{\sum_k d_{kd}}{U}, \quad \forall l,d,$$

where the capacity utilization U is 0.4, 0.6 or 0.8. It is important to notice that the utilization of capacity is only an estimate, as setup times do not influence the computation of Cap_{ld}. In summary we have:

Table 1: Size of the instances in terms of the number of binary variables, number of continuous variables and number of constraints

J	15			10			5		
D	20	10	5	20	10	5	20	10	5
Number of binaries	9,000	4,500	2,250	4,000	2,000	1,000	1,000	500	250
Number of continuous	75,151	36,826	18,226	24,101	11,551	5,651	4,051	1,776	826
Number of constraints	23,405	11,695	5,840	10,610	5,300	2,645	2,815	1,405	700

$$|5,10,15| \times |5,10,20| \times |50| \times |0.4,0.6,0.8| = 27 \text{ instances.}$$

In Table 1 the size of each instance that is part of the computational study is scrutinized in terms of variables and constraints of the MIP formulation.

5.2 Evaluation Metrics

In order to be able to evaluate the quality of the non-dominated Pareto-front computed with the introduction of the dual mutation operator, we rely on two performance measures: one unary metric, the hypervolume [32] and the first-order empirical attainment functions [23].

The hypervolume metric represents the volume of the objective space that is dominated by a solution set. To calculate this value a reference point is needed. For minimization problems, its value is set to exceed the maximal values for each objective. Hence, in our experiments it is found automatically with the best extreme points for each objective found for each instance. When using this metric to compare the performance of two or more solution methods, the one giving solutions with the largest covered hypervolume is considered the best.

In addition, the first-order empirical attainment functions (EAFs) are used to represent the probabilistic performance of MOEAs by measuring the attainment of a reference set based on the generated solutions using multiple runs. Therefore, different curves are plotted, such as the minimum attainment surface and the maximum attainment surface, based on all runs. We also plot an average attainment surface representing the surface which was attained in 50% of the runs.

5.3 Parameter Tuning

The setting of the parameters is the following: 100 generations, 100 individuals, a crossover rate of 0.9 and a mutation rate (for standard and dual mutation) of $\frac{2}{D \cdot J}$. These values were tuned through exhaustive preliminary tests and recommendations in the literature. The specific values of the dual mutation operator were set to: $\alpha = 0$, $\beta = 0.1$ and $\gamma = 1$.

The reason behind the value of $\alpha = 0$ was already given. Regarding the other values we should refer to the two objective functions. Clearly, the cost minimization is harder to solve than the freshness maximization due to the combinatorial nature of the former. Thus, we bring more weight on the dual mutation parameter that can improve this harder objective. It is important to note that if α, β and γ are set to the same values, the dual mutation is equal to the standard mutation operator since the probabilities to mutate each gene to any possible block are the same.

5.4 Experimental Results

To account for the randomness of the MOEA, 10 runs were executed for each instance. To illustrate the results for all instances, the hypervolume indicator and the differences between the first-order EAFs of the two algorithm variants are given for the instances with 60% capacity utilization in Figs. 3–8, respectively. The *Std* results refer to the variant where the standard mutation operator (cf. Sect. 4.2.1) acts on both feasible and infeasible solutions. The *Dual* results are obtained when the dual mutation operator is activated on feasible individuals (cf. Sect. 4.2.2), letting the standard mutation operator act on infeasible ones.

Note that the EAF plots have both objectives as minimization ones. Moreover, all figures have the instance identified in the form of *JxD*. In Table 2 median values for all instances with and without dual mutation are given. Moreover, the Kruskal–Wallis non-parametric statistical test is used for assessing if the hypervolume results differed significantly between algorithms and the respective *p-value* is computed (significant values are in bold).

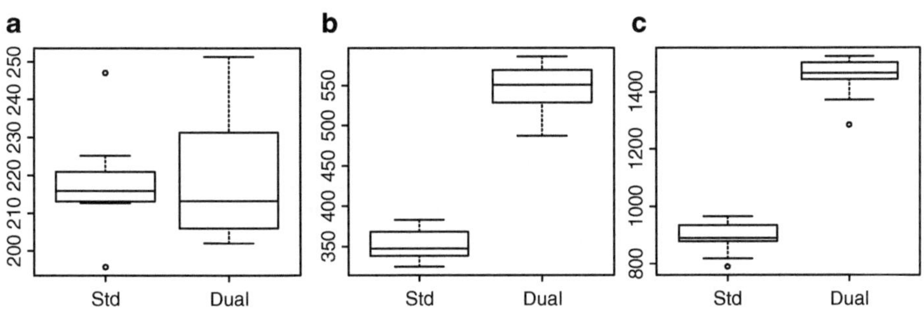

Fig. 3: Boxplots of the hypervolume indicator for instances with 5 products and 60% of capacity utilization (**a**) 5×5 (**b**) 5×10 (**c**) 5×20

First, it should be noticed that the algorithm is able to solve all instances of this difficult problem, even instances that could not be solved in the original paper where they were introduced. The running times of the algorithm increase with the number of blocks J and macro-periods D. Hence, it runs from 5 to 30 min in order to attain

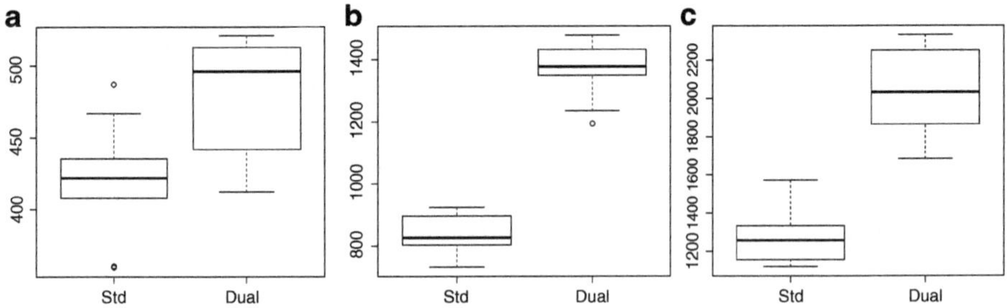

Fig. 4: Boxplots of the hypervolume indicator for instances with 10 products and 60% of capacity utilization (**a**) 10×5 (**b**) 10×10 (**c**) 10×20

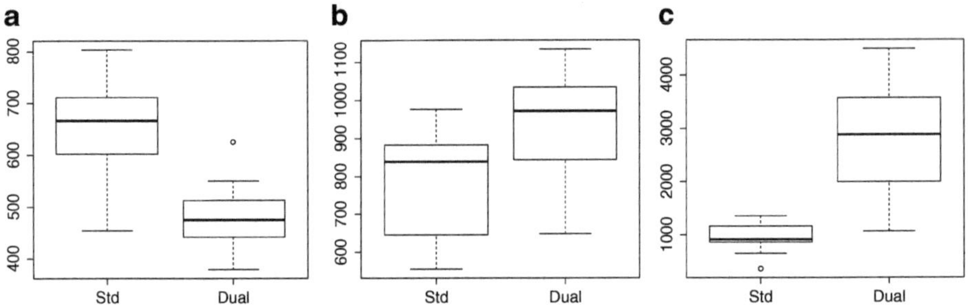

Fig. 5: Boxplots of the hypervolume indicator for instances with 15 products and 60% of capacity utilization (**a**) 15×5 (**b**) 15×10 (**c**) 15×20

Table 2: Hypervolume medians for the two versions of the MOEA and *p-values* of the Kruskal–Wallis statistical tests

		$U = 0.4$			$U = 0.6$			$U = 0.8$		
J	D	Std	Dual	*p-values*	Std	Dual	*p-values*	Std	Dual	*p-values*
5	5	174.62	250.96	**0.000**	215.92	213.21	0.762	347.52	550.71	**0.000**
5	10	572.81	825.31	**0.000**	347.52	550.71	**0.000**	621.66	632.20	0.496
5	20	1,038.66	1,874.71	**0.000**	889.47	1,464.54	**0.000**	1,247.58	1,306.87	0.013
10	5	457.94	440.88	0.406	422.00	496.04	**0.010**	616.32	535.42	**0.008**
10	10	1,043.31	952.26	**0.001**	826.38	1,377.33	**0.000**	616.32	1,066.71	**0.000**
10	20	1,535.19	2,092.29	**0.002**	1,257.10	2,031.76	**0.000**	1,676.61	1,426.47	**0.000**
15	5	511.58	580.56	0.112	665.88	474.35	**0.002**	357.38	180.45	**0.000**
15	10	982.60	812.58	0.013	838.96	972.30	0.023	752.55	579.73	0.034
15	20	516.54	589.03	**0.000**	914.79	2,884.26	**0.000**	516.54	589.03	**0.000**

100 generations. For higher values of the utilization coefficient U it becomes harder to spot feasible solutions. It should be highlighted that the difference between the running times of the MOEA with and without the dual mutation is only of a few seconds.

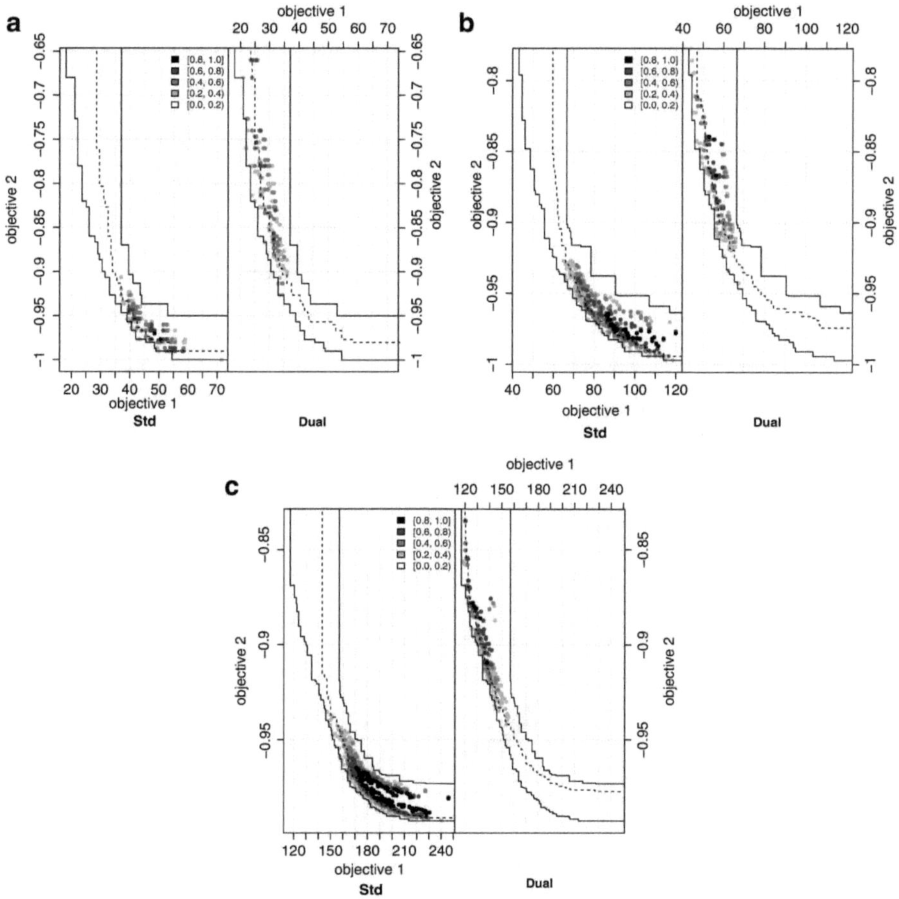

Fig. 6: EAF plots for instances with 5 products and 60% of capacity utilization (**a**) 5×5 (**b**) 5×10 (**c**) 5×20

The hypervolume indicator gives strong evidence that the dual operator is able to get better solutions in terms of Pareto-front optimality. This superiority seems to increase with the number of micro-periods. When setting the parameters of the dual operator it was referred that the effort was given upon the cost minimization objective. The EAF plots show that the dual operator is able to achieve indeed lower costs. On the other hand, the EAF plots evidence that the best solutions in terms of freshness are obtained by the MOEA without dual mutation, suggesting that β values could be increased to that end.

Figure 9 shows the impact of increasing β values in order to find solutions closer to the freshness optimum. Notice that the difference between the algorithm variants

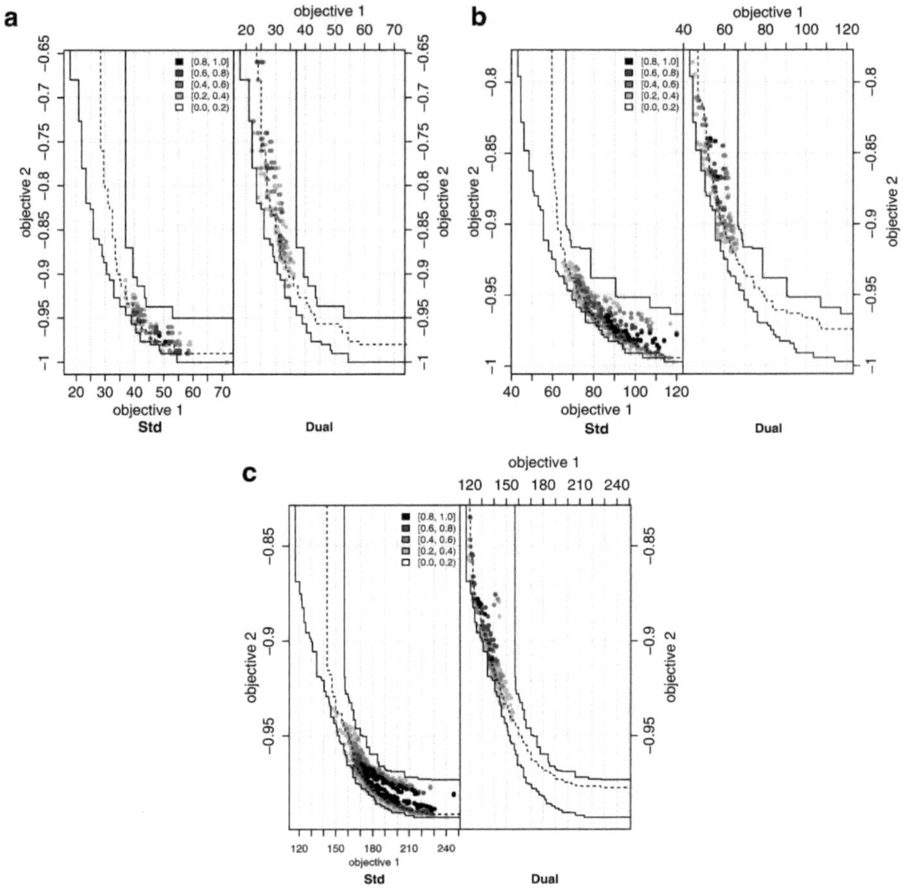

Fig. 7: EAF plots for instances with 10 products and 60% of capacity utilization (**a**) 10×5 (**b**) 10×10 (**c**) 10×20

with and without dual mutation is almost negligible for $\beta = 2$. However, in the variant with dual mutation, the average attainment surface is much closer to the maximum attainment surface.

6 Conclusions and Future Research

The main contribution of this paper lies in the exploration of hybrid methods to solve complex combinatorial multi-objective problems, such as the multi-objective lot-sizing and scheduling of perishable goods. Specifically, a new dual mutation operator is proposed to give tailored mutation probabilities to each individual based

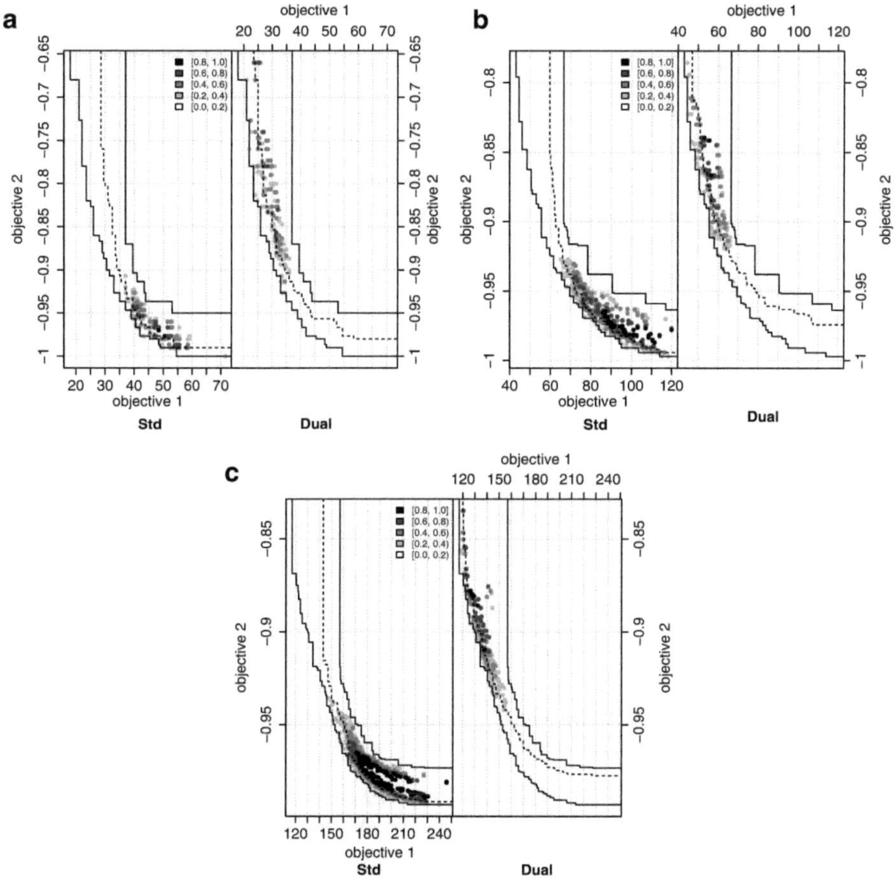

Fig. 8: EAF plots for instances with 15 products and 60% of capacity utilization (**a**) 15×5 (**b**) 15×10 (**c**) 15×20

on the reduced costs of the encoded variables coming from the LP relaxation. This operator enhances consistently the performance of the base MOEA through a better and directed exploration of the search space, ignoring mutation paths that do not seem promising. The rationale behind this dual mutation operator is replicable for other solution approaches in problems where the presence of LP relaxations makes possible to embed the use of dual values to guide the search.

Future work shall be devoted to the on-line adaptative setting of the dual mutation parameters and to testing this operator in multi-objective problems with a similar structure to the one presented in this paper. Moreover, the comparison of the performance in other MOEAs is of great interest.

Acknowledgements The first author appreciates the support of the FCT Project PTDC/EGE-GES/104443/2008 and the FCT Grant SFRH/BD/68808/2010.

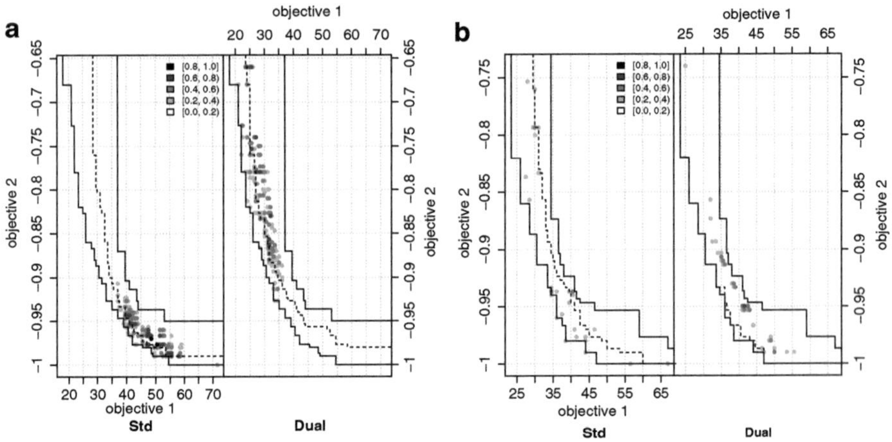

Fig. 9: EAF plots for illustrating the importance of dual mutation parameters in achieving different areas of the Pareto-front (**a**) 5×5 and U = 0.6 with $\beta = 0.1$ (**b**) 5×5 U = 0.6 with $\beta = 2$

References

1. Akkerman, R., Farahani, P., Grunow, M.: Quality, safety and sustainability in food distribution: a review of quantitative operations management approaches and challenges. OR Spectr. **32**(4), 863–904 (2010)
2. Almada-Lobo, B., Klabjan, D., Carravilla, M., Oliveira, J.: Single machine multi-product capacitated lot sizing with sequence-dependent setups. Int. J. Prod. Res. **45**(20), 4873–4894 (2007)
3. Amorim, P., Antunes, C., Almada-Lobo, B.: Multi-objective lot-sizing and scheduling dealing with perishability issues. Ind. Eng. Chem. Res. **50**(6), 3371–3381 (2011)
4. Amorim, P., Meyr, H., Almeder, C., Almada-Lobo, B.: Managing perishability in production-distribution planning: a discussion and review. Flexible Serv. Manuf. J. Doi: 10.1007/s10696-011-9122-3 published online
5. Chu, P., Beasley, J.: Constraint handling in genetic algorithms: the set partitioning problem. J. Heuristics **4**(4), 327–357 (1998)
6. Deb, K.: Multiobjective Optimization Using Evolutionary Algorithms. Wiley, Chichester (2001)
7. Deb, K., Pratap, A., Agarwal, S., Meyarivan, T.: A fast and elitist multiobjective genetic algorithm: NSGA-II. IEEE Trans. Evol. Comput. **6**(2), 182–197 (2002)
8. Ehrgott, M., Gandibleux, X.: Hybrid metaheuristics for multi-objective combinatorial optimization. In: Blum, C., Blesa Aguilera, M., Roli, A., Sampels, M. (eds.) Hybrid Metaheuristics: An Emerging Approach to Optimization, pp. 221–229. Springer, Berlin (2008)
9. Fleischmann, B., Meyr, H.: GLSP. OR Spectr. **19**, 11–21 (1997)
10. Gandibleux, X., Freville, A.: Tabu search based procedure for solving the 0-1 multiobjective knapsack problem: the two objectives case. J. Heuristics **6**, 361–383 (2000)
11. Gellert, T., Höhn, W., Möhring, R.: Sequencing and scheduling for filling lines in dairy production. Optim. Lett. **5**(3), 491–504 (2011)
12. Goldberg, D.E.: Genetic algorithms in search, optimization and machine learning. Addison-Wesley, Reading (1989)

13. Gomes da Silva, C., Figueira, J., Clímaco, J.: Integrating partial optimization with scatter search for solving bi-criteria {0,1}-knapsack problems. Eur. J. Oper. Res. **177**(3), 1656–1677 (2007)
14. Guimaraes, L., Klabjan, D., Almada-Lobo, B.: Annual production budget in the beverage industry. Eng. Appl. Artif. Intell. **25**(2), 229–241 (2012)
15. Günther, H.O., Grunow, M., Neuhaus, U.: Realizing block planning concepts in make-and-pack production using MILP modelling and SAP APO. Int. J. Prod. Res. **44**(18–19), 3711–3726 (2006)
16. Haase, K., Kimms, A.: Lot sizing and scheduling with sequence-dependent setup costs and times and efficient rescheduling opportunities. Int. J. Prod. Econ. **66**(2), 159–169 (2000)
17. Jozefowiez, N., Semet, F., Talbi, E.G.: The bi-objective covering tour problem. Comput. Oper. Res. **34**(7), 1929–1942 (2007)
18. Khare, V., Yao, X., Deb, K.: Performance Scaling of Multi-objective Evolutionary Algorithms. Springer, Berlin (2003)
19. Kopanos, G., Puigjaner, L., Georgiadis, M.: Optimal production scheduling and lot-sizing in dairy plants: the yogurt production line. Ind. Eng. Chem. Res. **49**(2), 701–718 (2010)
20. Kopanos, G., Puigjaner, L., Georgiadis, M.: Efficient mathematical frameworks for detailed production scheduling in food processing industries. Comput. Chem. Eng. **35**, 2929–2944 (2011)
21. Kopanos, G., Puigjaner, L., Georgiadis, M.: Production scheduling in multiproduct multistage semicontinuous food processes. Ind. Eng. Chem. Res. **50**(10), 6316–6324 (2011)
22. Kopanos, G., Puigjaner, L., Maravelias, C.: Production planning and scheduling of parallel continuous processes with product families. Ind. Eng. Chem. Res. **50**(3), 1369–1378 (2011)
23. López-Ibáñez, M., Paquete, L., Stützle, T.: Exploratory analysis of stochastic local search algorithms in biobjective optimization. In: Bartz-Beielstein, T., Chiarandini, M., Preuss, M. (eds.) Experimental Methods for the Analysis of Optimization Algorithms, pp. 209–222. Springer, Berlin (2010)
24. Lütke Entrup, M., Günther, H.O., Van Beek, P., Grunow, M., Seiler, T.: Mixed-integer linear programming approaches to shelf-life-integrated planning and scheduling in yoghurt production. Int. J. Prod. Res. **43**(23), 5071–5100 (2005)
25. Marinelli, F., Nenni, M., Sforza, A.: Capacitated lot sizing and scheduling with parallel machines and shared buffers: a case study in a packaging company. Ann. Oper. Res. **150**(1), 177–192 (2007)
26. Meyr, H.: Simultaneous lotsizing and scheduling on parallel machines. Eur. J. Oper. Res. **139**(2), 277–292 (2002)
27. Oliveira Santos, M., Massago, S., Almada-Lobo, B.: Infeasibility handling in genetic algorithm using nested domains for production planning. Comput. Oper. Res. **37**(6), 1113–1122 (2010)
28. Pahl, J., Voß, S.: Discrete lot-sizing and scheduling including deterioration and perishability constraints. In: Dangelmaier, W., Blecken, A., Delius, R., Klöpfer, S. (eds.) Advanced Manufacturing and Sustainable Logistics. Lecture Notes in Business Information Processing, vol. 46, pp. 345–357. Springer, Berlin (2010)
29. Pahl, J., Voß, S., Woodruff, D.: Discrete lot-sizing and scheduling with sequence-dependent setup times and costs including deterioration and perishability constraints. In: IEEE, 44th Hawaii International Conference on System Sciences (2011)
30. Ponnambalam, S., Reddy, M.: A GA-SA multiobjective hybrid search algorithm for integrating lot sizing and sequencing in flow-line scheduling. Int. J. Adv. Manuf. Technol. **21**, 126–137 (2003)
31. Przybylski, A., Gandibleux, X., Ehrgott, M.: Two phase algorithms for the bi-objective assignment problem. Eur. J. Oper. Res. **185**(2), 509–533 (2008)
32. Zitzler, E., Thiele, L.: Multiobjective evolutionary algorithms: a comparative case study and the strength Pareto approach. IEEE Trans. Evol. Comput. **3**(4), 257–271 (1999)

Brain Cine-MRI Registration Using MLSDO Dynamic Optimization Algorithm

Julien Lepagnot, Amir Nakib, Hamouche Oulhadj, and Patrick Siarry

Abstract In this chapter, we propose to use a dynamic optimization algorithm to assess the deformations of the wall of the third cerebral ventricle in the case of a brain cine-MRI. In this method, a segmentation process is applied to a 2D+t cine-MRI sequence to detect the contours of a region of interest (i.e. lamina terminalis). Then, successive segmented contours are matched using a global alignment procedure, followed by a registration process. This registration process consists in optimizing an objective function that can be considered as a dynamic function. Thus, a dynamic optimization algorithm, called MLSDO, is used to solve the registration problem. The results obtained by MLSDO are compared to those of several well-known static optimization algorithms. This comparison shows the efficiency of MLSDO, and the relevance of using a dynamic optimization algorithm to solve this kind of problems.

1 Introduction

Image registration is the process of overlaying two or more images of the same scene taken at different times, from different viewpoints, and/or by different sensors. It is a critical step in all image analysis tasks in which the final information is gained from the combination of various data sources like, in image fusion or change detection.

It geometrically aligns two images: the source and the target images. It is done by determining a transformation that maps the target image to the source one. Thus, registering a sequence of images consists in determining, for each couple of successive images, the transformation that makes the first image of the couple match the following image.

Julien Lepagnot · Amir Nakib · Hamouche Oulhadj · Patrick Siarry
LISSI, Université Paris-Est Créteil (UPEC), Créteil, France
e-mail: julien.lepagnot, amir.nakib, oulhadj, siarry@u-pec.fr

L. Di Gaspero et al. (eds.), *Advances in Metaheuristics*, Operations Research/ Computer Science Interfaces Series 53, DOI 10.1007/978-1-4614-6322-1_6,

If successive images of the sequence are correlated, then we can make use of this correlation to accelerate the registration process. This can be done by using a *dynamic optimization* algorithm.

Recently, optimization in dynamic environments has attracted a growing interest, due to its practical relevance. Almost all real-world problems are time dependent or dynamic, i.e. their objective function changes over the time. For dynamic environments, the goal is not only to locate the global optimum, but also to track it as closely as possible over the time. Then, a dynamic optimization problem can be expressed as in Eq. (1), where $f(\mathbf{x},t)$ is the objective function of a minimization problem, $h_j(\mathbf{x},t)$ denotes the jth equality constraint among u equality constraints and $g_k(\mathbf{x},t)$ denotes the kth inequality constraint among v inequality constraints. All of these functions may change over the time (iterations), as indicated by the dependence on the time variable t.

$$\begin{aligned} \min\ & f(\mathbf{x},t) \\ \text{s.t.}\ & h_j(\mathbf{x},t) = 0 \text{ for } j = 1,2,\ldots,u \\ & g_k(\mathbf{x},t) \leq 0 \text{ for } k = 1,2,\ldots,v \end{aligned} \tag{1}$$

In this work, we focus on a dynamic optimization problem with time constant constraints. We propose to apply the *Multiple Local Search algorithm for Dynamic Optimization* (MLSDO) [8] to the registration of sequences of segmented images.

Hydrocephalus pathology consists in an abnormal accumulation of cerebrospinal fluid in the ventricles, or cavities, of the brain. This may cause increased intracranial pressure inside the skull and progressive enlargement of the head, convulsion, tunnel vision, and mental disability. Hydrocephalus may be suggested by symptoms; however, imaging studies of the brain are the mainstay of diagnosis. We focus on a method based on cine-MRI sequences to facilitate this diagnosis, and to assist neurosurgeons in the characterization of the pathology at hand [12].

In the problem we deal with, each image contains segmented contours of the region of interest (i.e. lamina terminalis), extracted from a brain cine-MRI sequence of 20 images. This sequence corresponds to 80% of a R-R cardiac cycle [12], and the segmentation method used to generate these contours is described in [12]. An example of two images extracted from a brain cine-MRI sequence is presented in Fig. 1. Hence, every segmented sequence is composed of 20 segmented cine-MRI. An example of segmented sequence is illustrated in Fig. 2. The goal is to match the points of the contours in each couple of successive images of this sequence. Transformations that result from this matching operation can then be used to assess the deformation movements of the third cerebral ventricle.

Several papers are proposed in the literature about the analysis and quantification of cardiac movements, we can cite those recently published [3, 2, 14]. In our case, the single paper that deals with the problem at hand is [12]. The main difference between the problem at hand and the cardiac problem lies in the amplitude of the movements of the ventricles. Indeed, the amplitude of the cardiac ventricle movements is higher than the amplitude of the cerebral ventricle movements. Then, a more accurate method is required in the cerebral case. In this chapter, we propose

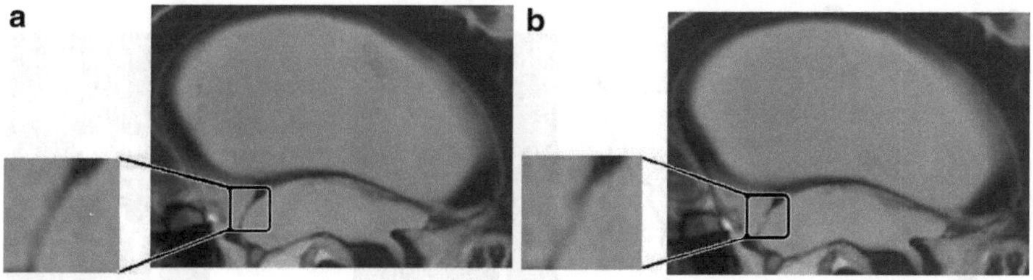

Fig. 1: Two images from a brain cine-MRI sequence: (**a**) first image of the sequence, (**b**) sixth image of the sequence

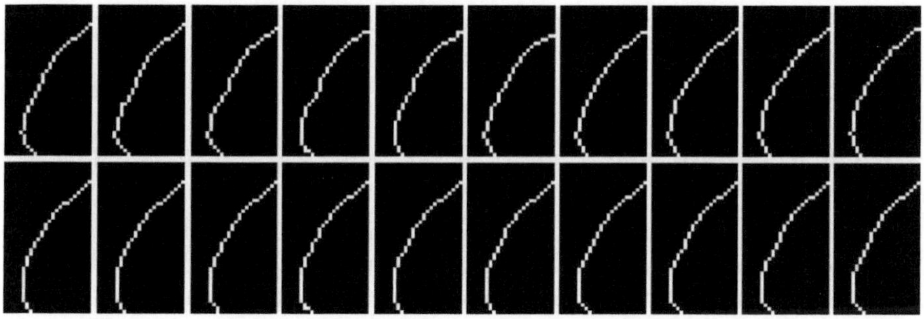

Fig. 2: Sequence of segmented images

a method inspired from [12] to assess the movements of a region of interest (ROI) from the wall of the third ventricle. The main contribution of the present work is to show the importance of the use of dynamic optimization algorithms for brain cine-MRI registration.

The rest of this chapter is organized as follows. In Sect. 2, the method proposed to register sequences of segmented contours is described. In Sect. 3, the MLSDO algorithm and its use in the problem at hand are presented. In Sect. 4, a comparison of the results obtained by MLSDO on this problem to the ones of several well-known static optimization algorithms is performed. This comparison shows the relevance of using MLSDO on this problem. Finally, a conclusion and the works under progress are given in Sect. 5.

2 The Registration Method

A method inspired from [12] is proposed in this chapter to register sequences of segmented contours. This operation is required to track the position of points belonging to the contours of the ROI over time (images of a sequence). This operation is carried out in two steps: a matching step and a registration process.

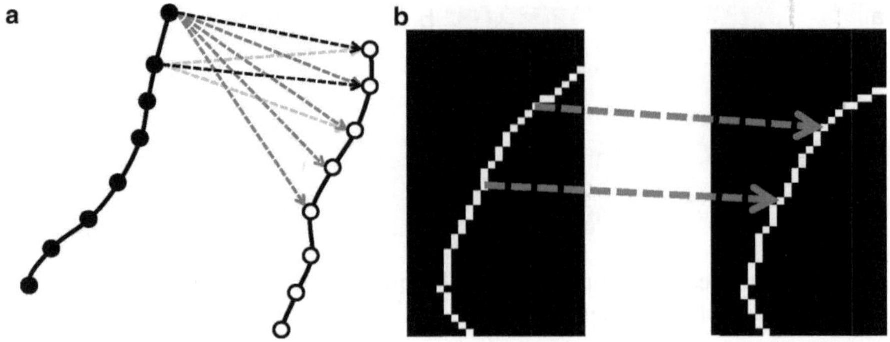

Fig. 3: Illustration of the matching procedure: (**a**) synthetic contours, (**b**) real contours of ROI (lamina terminalis) [12]

2.1 The Matching Step

An example is presented in Fig. 3 to illustrate this step. In Fig. 3a we present two synthetic contours to be aligned, the goal is to associate each point of the initial contours to one point of the second contours. We assume that each point of the first contours can be associated to at least one point of the second contours. Figure 3b illustrates the goal of this procedure on real contours.

Let C_1 and C_2 be two curves consisting of L_1 and L_2 points, respectively, and corresponding to two successive contours of the sequence. Then, at the end of this alignment step, we get a set C_1' of sorted points of the contours C_2 that correspond to the contours C_1.

2.2 The Registration Process

In order to evaluate the pulsatile movements of the third cerebral ventricle in a way that it can be easily interpreted, a rigid transformation is used. We assume that a transformation T_Φ allows $C_2 = T_\Phi(C_1)$ and, for every point $(x_1\ y_1)^{\mathrm{T}}$ of C_1 (the transpose of a matrix A is denoted by A^{T}), it is defined by:

$$\begin{pmatrix} x_2 \\ y_2 \end{pmatrix} = \begin{pmatrix} s_1.\cos\theta & -s_2.\sin\theta \\ s_1.\sin\theta & s_2.\cos\theta \end{pmatrix} \cdot \begin{pmatrix} x_1 \\ y_1 \end{pmatrix} + \begin{pmatrix} t_x \\ t_y \end{pmatrix} \tag{2}$$

Using homogeneous coordinates, Eq. (2) becomes:

$$\begin{pmatrix} x_2 \\ y_2 \\ 1 \end{pmatrix} = \begin{pmatrix} s_1.\cos\theta & -s_2.\sin\theta & t_x \\ s_1.\sin\theta & s_2.\cos\theta & t_y \\ 0 & 0 & 1 \end{pmatrix} \cdot \begin{pmatrix} x_1 \\ y_1 \\ 1 \end{pmatrix} \tag{3}$$

The parameters s_1 and s_2 correspond to a homothetic transformation, θ corresponds to a rotation, t_x and t_y correspond to a translation operation. The set of parameters $\boldsymbol{\Phi} = \{s_1, s_2, \theta, t_x, t_y\}$ is estimated through the minimization of the mean square error (MSE) computed on the set of points that will be paired. Then, the optimization criterion is given by:

$$MSE(\boldsymbol{\Phi}) = \sum_{j=1}^{L_1} \frac{[C_1'(j) - T_{\boldsymbol{\Phi}}(C_1(j))]^{\mathrm{T}} [C_1'(j) - T_{\boldsymbol{\Phi}}(C_1(j))]}{L_1} \tag{4}$$

where $C_1'(j)$ and $C_1(j)$ denote the jth point of C_1' and the jth point of C_1, respectively.

Then, the registration problem can be formulated as an optimization problem, and the optimal set of parameters $\boldsymbol{\Phi}^*$ can be found by solving the following problem:

$$\boldsymbol{\Phi}^* = \text{Min}\, MSE(\boldsymbol{\Phi}) \tag{5}$$

Then, to avoid being trapped in a local optimum, a metaheuristic is applied on this problem. The presence of local optima in the objective function is indeed a common and serious problem for registration methods [6]. For the problem at hand, a recently proposed dynamic optimization algorithm, MLSDO, is used.

3 The MLSDO Algorithm

In this section, MLSDO and its use on the problem at hand are described. At first, the algorithm is presented. Then, the dynamic objective function proposed for the problem at hand is described. Afterwards, the parameter fitting of MLSDO is given to solve this problem.

3.1 Description of the Algorithm

MLSDO uses several local searches, each one performed in parallel with the others, to explore the search space, and to track the found optima over the changes in the objective function. Each local search is performed by an agent, and all the agents are coordinated by a dedicated module (the coordinator). Two types of agents exist in MLSDO: the exploring agents (to explore the search space in order to discover the local optima), and the tracking agents (to track the found local optima over the changes in the objective function). The local searches performed by the exploring agents have a greater initial step size than the one of the tracking agents, because the exploring agents have to widely explore the search space. The strategies used to coordinate these local search agents enable the fast convergence to well diversified optima, to react quickly to a change and to find the global optimum. Especially, each agent performs its local search in an exclusive area of the search space: an exclusion

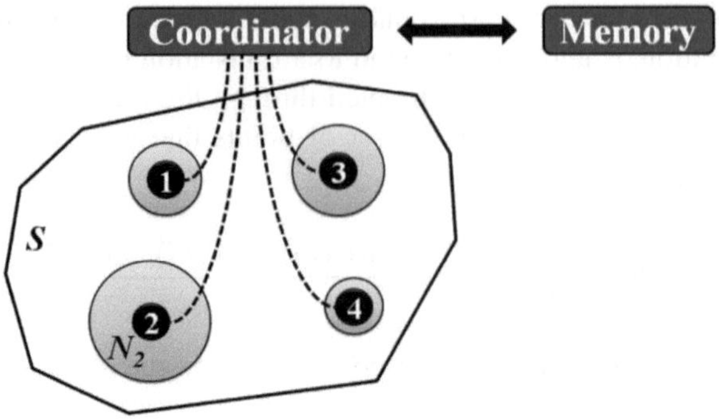

Fig. 4: Overall scheme of MLSDO

radius is attributed to each agent. This way, if several agents converge to a same local optimum, then only one of them can continue to converge to this local optimum: all the other conflicting agents are reinitialized elsewhere in the search space. Another important strategy is the use of two levels of precision in the stopping criterion of the local searches of the agents. In this way, we prevent the fine-tuning of low quality solutions, which could lead to a waste of fitness function evaluations; only the best solution found by MLSDO is fine-tuned. Furthermore, the local optima found during the optimization process are archived, to accelerate the detection of the global optimum after a change in the objective function. These archived optima are used as initial solutions of the local searches performed by the tracking agents. The overall scheme of MLSDO is illustrated in Fig. 4, where the local search agents are depicted by the numbered black-filled circles in the search space S, and the neighborhood of the ith agent is denoted by N_i.

The flowchart of the search procedure of an agent is illustrated in Fig. 5. One can see that two special states, named "SYNCHRONIZATION A" and "SYNCHRONIZATION B," appear in this flowchart. These states mark the end of one step of the procedure of an agent. Hence, if one of these states has been reached, then the agent halts its execution until all other agents have reached one of these states. Afterwards, the execution of the agents is resumed; i.e., if an agent halts on SYNCHRONIZATION A (SYNCHRONIZATION B, respectively), then it resumes its execution on SYNCHRONIZATION A (SYNCHRONIZATION B, respectively). This special state allows the parallel execution of the agents. More details about this algorithm are in [7].

MLSDO has been compared to other leading dynamic optimization algorithms using two of the main benchmarks available in the literature : the Moving Peaks Benchmark (MPB) [1] and the Generalized Dynamic Benchmark Generator (GDBG) [10, 11]. The benchmark results show the efficiency of MLSDO [7, 9]:

- MLSDO is ranked at the first place among seven algorithms tested using GDBG ;
- MLSDO is ranked at the second place among 16 algorithms tested using MPB.

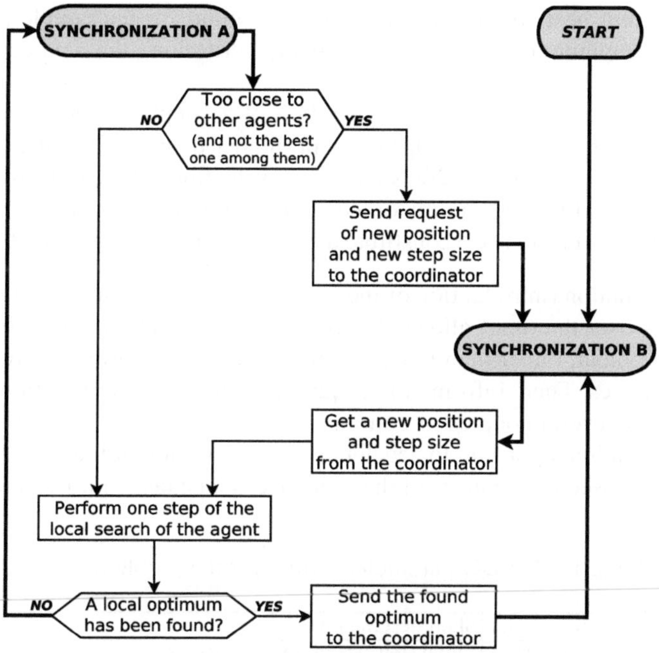

Fig. 5: Flowchart of the main procedure of a MLSDO agent

3.2 Cine-MRI Registration as a Dynamic Optimization Problem

The registration of a cine-MRI sequence can be seen as a dynamic optimization problem. Then, the dynamic objective function optimized by MLSDO changes according to the following rules:

- The criterion in Eq. (4) has to be minimized for each couple of contours, as we are in the case of a sequence, then the optimization criterion becomes:

$$MSE(\Phi(t),t) = \sum_{j=1}^{L_t} \frac{\left[C'_t(j) - T_{\Phi(t)}(C_t(j))\right]^{\mathrm{T}} \left[C'_t(j) - T_{\Phi(t)}(C_t(j))\right]}{L_t} \quad (6)$$

  where t is the index of the contours on which the transformation $T_{\Phi(t)}$ is applied, also equal to the index of the current couple of contours in the sequence. $\Phi(t)$, $C'_t(j)$, $C_t(j)$ and L_t are the same as Φ, C'_1, C_1 and L_1 defined before, respectively, but here are dependent on the couple of contours.
- Then, the dynamic optimization problem is defined by:

$$\text{Min } MSE(\Phi(t),t) \quad (7)$$

- If the current best solution (transformation) found for the couple t cannot be improved anymore (according to a stagnation criterion), the next couple $(t+1)$ is treated.
- The stagnation criterion of the registration of a couple of successive contours is satisfied if no significant improvement (higher than 1E-5) in the current best solution is observed during 200 successive evaluations of the objective function.
- Thus, the end of the registration of a couple of contours and the beginning of the registration of the next one constitute a change in the objective function.

This formulation (introduction of the time variable t to get an objective function that changes over the time) allows the use of dynamic optimization algorithms to solve this problem, rather than having to restart a static algorithm to register a sequence of images. Then, information acquired on the objective function during the registration of several couples of contours, in a sequence, can be used by the dynamic optimization algorithm to accelerate the registration of the next couples (the correlations between the images of the sequence can be taken into account).

Table 1: MLSDO parameter setting for the problem at hand

Name	Type	Interval	Value	Short description
r_l	Real	$(0, r_e)$	1E-3	Initial step size of tracking agents
r_e	Real	$(0, 1]$	1E-1	Exclusion radius of the agents, and initial step size of exploring agents
δ_{ph}	Real	$[0, \delta_{pl}]$	1E-7	Highest precision parameter of the stopping criterion of the agents local searches
δ_{pl}	Real	$[\delta_{ph}, +\infty]$	1E-5	Lowest precision parameter of the stopping criterion of the agents local searches
n_a	Integer	$[1, 10]$	1	Maximum number of exploring agents
n_c	Integer	$[0, 20]$	2	Maximum number of tracking agents created after the detection of a change

3.3 Parameter Fitting of MLSDO

Table 1 summarizes the six parameters of MLSDO that the user has to define. These values will be used to perform the experiments reported in the following section.

In this table, the values given are suitable for the problem at hand, and they were fixed experimentally. Among several sets of values for the parameters, we selected the one that minimizes the number of evaluations performed. More precisely, each parameter has been fitted manually, one after the other, since there is no important correlation between them [7]. Besides, the obtained set of fitted parameters is optimal for each sequence in our data set. One can see that only one exploring agent is used to solve this problem. It is indeed sufficient for this problem, and using more than one exploring agent increases the number of evaluations required to register a sequence. However, using more than one exploring agent can improve the performance of MLSDO on other problems.

4 Experimental Results and Discussion

A comparison between the results obtained by MLSDO and those obtained by several well-known static optimization algorithms is presented in this section. These algorithms, and their parameter setting, empirically fitted to the problem at hand, are defined below (see references for more details on these algorithms and their parameter fitting). Since these algorithms have a low number of parameters, they can be fitted by a thorough exploration of their possible values, in order to get the optimal set of parameters. As for MLSDO, this optimal set is independent of the cine-MRI sequence used:

- CMA-ES (*Covariance Matrix Adaptation Evolution Strategy*) [5] using the recommended parameter setting, except for the initial step size σ, set to $\sigma = 0.3$. The population size λ and the number of selected individuals μ are set to $\lambda = 8$ and $\mu = 4$. This algorithm is used in the optimization process of the paper from which the registration method proposed in this chapter is inspired [12] ;
- SPSO-07 (*Standard Particle Swarm Optimization* in its 2007 version) [4] using the recommended parameter setting, except for the number S of particles ($S = 6$) and for the parameter K used to generate the particles neighborhood ($K = 5$) ;
- DE (*Differential Evolution*) [13] using the "DE/target-to-best/1/bin" strategy, a number of parents equal to $NP = 12$, a weighting factor $F = 0.8$, and a crossover constant $CR = 0.9$.

As these algorithms are static, we have to consider the registration of each couple of successive contours as a new problem to optimize. Thus, these algorithms are restarted after the registration of each couple of contours, using the stagnation criterion defined in Sect. 3.2. The results obtained using MLSDO, as a static optimization algorithm, are also given.

The static optimization algorithms should be restarted for each couple of contours, in order to restore their diversification abilities, and to avoid the problem of outdated memory (the recorded values of the dynamic objective function may vary after each change in the function). However, we tried to accelerate their convergence using a simple strategy. Rather than restarting the algorithms from random solutions, the best solution found for the last couple of contours is used as a starting solution for the next couple. Yet, this strategy does not lead to any improvement in the performance of the registration process, and it can even worsen the results. Indeed, using this strategy, SPSO-07 and DE are unable to reach satisfying solutions. CMA-ES is able to reach them and, thus, to register the sequences. However, we found no significant difference in the number of evaluations performed by CMA-ES with

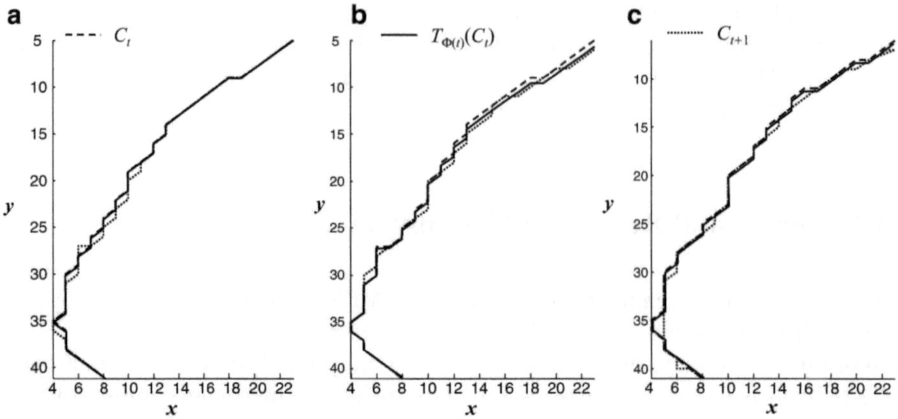

Fig. 6: Illustration of the registration of the first three couples of contours of a sequence: (**a**) first couple of contours, (**b**) second couple of contours, (**c**) third couple of contours

Table 2: Average number of evaluations and execution time to register all couples of contours, and average sum of MSE, obtained by each algorithm

	Algorithm	Evaluations	Execution time (min)	Cumulated MSE
Dynamic optimization	MLSDO	$13,841.80 \pm 386.55$	13.2 ± 5.7	$2.05 \pm$ 1.3E-4
Static optimization	CMA-ES	$17,651.65 \pm 288.10$	$5,029.5 \pm 212.8$	$2.05 \pm$ 1.6E-2
	SPSO-07	$19,611.70 \pm 994.13$	49.8 ± 22.5	$2.05 \pm$ 2.7E-2
	DE	$21,749.60 \pm 869.46$	56.4 ± 23.4	$2.05 \pm$ 2.8E-5
	MLSDO	$28,891.15 \pm 908.88$	29.8 ± 20.3	$2.05 \pm$ 1.8E-4

and without this strategy, using a Wilcoxon–Mann–Whitney statistical test at a 99% confidence level.

The registrations of the first three couples of contours, performed by MLSDO, are illustrated in Fig. 6. In this figure, we can see that the use of a rigid transformation in the registration process does not lead to perfect results. However, it provides an easily interpretable model of the movements in the ROI.

In Table 2, the average number of evaluations among 20 runs of the algorithms are given. The average execution time of each algorithm, written in C, among 20 runs and using a 2.27 GHz Intel Core i5 processor, is also given. Additionally, the sum of the mean square errors [see Eqs. (4) and (6)] of each registration of the sequence is given in this table, averaged on 20 runs of the algorithms. The convergence curve of MLSDO, and that of the best performing static optimization algorithm on the problem at hand, i.e. CMA-ES, are shown in Fig. 7. In this figure, the number of evaluations per registration of a couple of contours is fixed to 2,000, in order to enable the comparison of the convergence of the algorithms. For readability, a logarithmic scale is used on the ordinates.

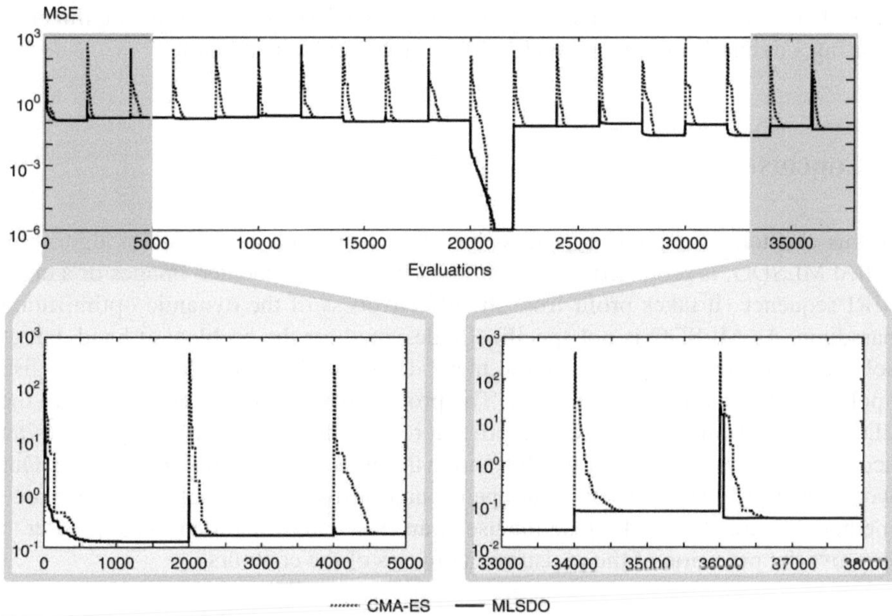

Fig. 7: Convergence graph of MLSDO and CMA-ES on the problem at hand

At first, the transformations obtained by all algorithms are not significantly different: the values found for the parameters s_1, s_2, θ, t_x and t_y do not differ more than 5.9E-3, 1.8E-3, 3.0E-3, 3.7E-3 and 4.5E-3, respectively. Especially, the mean square error of the registration of a couple of contours does not differ more than 8.5E-5 between two algorithms. This level of precision is satisfying for the problem at hand. Then, the average sum of MSE given in Table 2 shows that the algorithms have a similar average precision.

The similar MSE values found by the different algorithms show that all of them provide solutions with a similar accuracy. Therefore, since similar quality of registration results are obtained, we can compare the performance of these algorithms in terms of number of evaluations required to register a whole sequence.

We can see in Table 2 that the number of evaluations of the objective function performed by MLSDO, used as a dynamic optimization algorithm, is significantly lower than the ones of the static optimization algorithms. A Wilcoxon–Mann–Whitney statistical test has been applied on the numbers of evaluations performed by MLSDO and CMA-ES, the best ranked algorithms in terms of number of evaluations. This test confirms at a 99% confidence level that there is a significant difference between their performances. It can be seen also in Fig. 7 that the convergence of MLSDO to an acceptable solution is faster than CMA-ES for the registration of all the couples of contours, except for the first one. MLSDO needs indeed to learn from the first registration in order to accelerate its convergence on the next ones. Thus, this comparison shows the efficiency of MLSDO and the significance of using a dynamic optimization algorithm on the problem at hand.

5 Conclusion

In this chapter, a registration process based on a dynamic optimization algorithm, called MLSDO, is proposed to register quickly all the segmented images of a cine-MRI sequence. It takes profit from the effectiveness of the dynamic optimization paradigm. As MLSDO is not specifically designed for the problem at hand, but to solve a wide range of problems, the method proposed in this chapter can be also applied to other kinds of sequences. The process is sequentially applied on all the 2D segmented images. The entire procedure is fully automated and provides an accurate assessment of the ROI deformation throughout the entire cardiac cycle. Our work under progress consists in an open source software encapsuling the proposed method. We are also working on the use of an elastic deformation model in order to improve the precision of the registration process of the contours.

References

1. Branke, J.: The Moving Peaks Benchmark website. http://www.aifb.uni-karlsruhe.de/~jbr/MovPeaks (1999)
2. Budoff, M.J., Ahmadi, N., Sarraf, G., Gao, Y., Chow, D., Flores, F., Mao, S.S.: Determination of left ventricular mass on cardiac computed tomographic angiography. Acad. Radiol. **16**(6), 726–732 (2009)
3. Chenoune, Y., Delchelle, E., Petit, E., Goissen, T., Garot, J., Rahmouni, A.: Segmentation of cardiac cine-MR images and myocardial deformation assessment using level set methods. Comput. Med. Imaging Graph. **29**(8), 607–616 (2005)
4. Clerc, M., et al.: The *Particle Swarm Central* website. http://www.particleswarm.info
5. Hansen, N., Ostermeier, A.: Completely derandomized self-adaptation in evolution strategies. Evol. Comput. **9**(2), 159–195 (2001)
6. Jenkinson, M., Bannister, P., Brady, M., Smith, S.: Improved optimization for the robust and accurate linear registration and motion correction of brain images. NeuroImage **17**(2), 825–841 (2002)
7. Lepagnot, J., Nakib, A., Oulhadj, H., Siarry, P.: A multiple local search algorithm for continuous dynamic optimization. (Under submission)
8. Lepagnot, J., Nakib, A., Oulhadj, H., Siarry, P.: Brain cine-MRI registration using MLSDO dynamic optimization algorithm. In: Proceedings of the 9th Metaheuristics International Conference (MIC 2011), vol. 1, pp. 241–249. Udine, Italy (2011)
9. Lepagnot, J., Nakib, A., Oulhadj, H., Siarry, P.: A dynamic multi-agent algorithm applied to challenging benchmark problems. In: Proceedings of IEEE Congress on Evolutionary Computation, pp. 1–8, IEEE Press Piscataway, NJ (2012)
10. Li, C., Yang, S.: A generalized approach to construct benchmark problems for dynamic optimization. In: Proceedings of the 7th International Conference on Simulated Evolution and Learning, pp. 391–400. Springer, Melbourne, Australia (2008)

11. Li, C., Yang, S., Nguyen, T.T., Yu, E.L., Yao, X., Jin, Y., Beyer, H.G., Suganthan, P.N.: Benchmark generator for CEC 2009 competition on dynamic optimization. Technical report, University of Leicester, University of Birmingham, Nanyang Technological University (2008)
12. Nakib, A., Aiboud, F., Hodel, J., Siarry, P., Decq, P.: Third brain ventricle deformation analysis using fractional differentiation and evolution strategy in brain cine-MRI. In: Medical Imaging 2010: Image Processing, vol. 7623, pp. 76, 232I–1–76,232I–10. SPIE, San Diego, California, USA (2010)
13. Price, K., Storn, R., Lampinen, J.: Differential Evolution - A Practical Approach to Global Optimization. Springer, Berlin (2005)
14. Sundar, H., Litt, H., Shen, D.: Estimating myocardial motion by 4D image warping. Pattern Recognit. **42**(11), 2514–2526 (2009)

GRASP with Path Relinking for the Two-Echelon Vehicle Routing Problem

Teodor Gabriel Crainic, Simona Mancini, Guido Perboli, and Roberto Tadei

Abstract We propose a meta-heuristic based on greedy randomized adaptive search procedure (GRASP) combined with path relinking to address the two-echelon vehicle routing problem (2E-VRP), an extension of the capacitated vehicle routing problem (CVRP) in which the delivery from a single depot to customers is achieved by routing and consolidating the freight through intermediate depots called satellites. The problem is treated by separating the depot-to-satellite transfer and the satellite-to-customer delivery, and iteratively solving the two resulting routing subproblems, while adjusting the satellite workloads that link them. The meta-heuristic scheme consists of applying a GRASP and a local search (LS) procedures in sequence. Then, the resulting solution is linked to an elite solution by means of a path relinking procedure. To escape from infeasible solutions, which are quite common in this kind of problem, a feasibility search (FS) procedure is applied within path relinking. Extensive computational results on instances with up to 50 customers and 5 satellites show that the meta-heuristic is able to improve literature results, both in efficiency and accuracy.

Teodor Gabriel Crainic
CIRRELT, Montreal, QC, Canada
e-mail: TeodorGabriel.Crainic@cirrelt.ca

Simona Mancini · Roberto Tadei
Politecnico di Torino, Turin, Italy
e-mail: simona.mancini@polito.it, roberto.tadei@polito.it

Guido Perboli
Politecnico di Torino, Turin, Italy
CIRRELT, Montreal, QC, Canada
e-mail: guido.perboli@polito.it

L. Di Gaspero et al. (eds.), *Advances in Metaheuristics*, Operations Research/ Computer Science Interfaces Series 53, DOI 10.1007/978-1-4614-6322-1_7,

1 Introduction

The aim of this paper is to present an efficient meta-heuristic to address the *two-echelon vehicle routing problem* (*2E-VRP*), named GRASP-PR. The *2E-VRP* is a variant of the capacitated vehicle routing problem (CVRP), characterized by a single depot and a given number of secondary facilities named satellites. The first level routing problem addresses depot-to-satellite delivery, while the satellite-to-customer delivery routes are built at the second level. The goal is to ensure an efficient and efficient operation of the system, where the demand is satisfied and the total cost of the traffic within the overall transportation network is minimized.

This problem is frequently faced in real-life applications, both at the strategic and tactical planning levels, and in day-to-day operations. Methods that can be applied at both levels must be accurate and fast. Thus, for planning, the *2E-VRP* is usually part of larger optimization frameworks, meaning that it must be solved many times during the optimization process, and computational times need to be limited. On the other hand, good feasible solutions are needed in a very short time when optimization problems are to be used at the operational level. Solution quality is crucial in all cases, because it directly impacts the revenues and service quality of the transportation company.

The meta-heuristic introduced in this paper is based on hybridizing greedy randomized adaptive search procedure (GRASP) and path relinking. More precisely, GRASP is used to generate solutions, which are post-optimized by means of a local search (LS) procedure. In order to improve the solution quality, a path between the current solution obtained by GRASP with the LS procedure and the best solution found so far is built by means of a path relinking procedure. The meta-heuristic is tested on medium-sized instances with 50 customers and 5 satellites, showing that the new method is able to improve existing state-of-the-art results.

The paper is organized as follows. The *2E-VRP* and the main literature results are presented in Sect. 2. Section 3 is dedicated to the GRASP-PR general framework, while computational tests and result analysis are reported in Sect. 4. Conclusions are drawn in Sect. 5.

2 Problem Definition and Literature Review

In the *2E-VRP*, the distribution of freight cannot be managed by direct shipping from the depot to the customers. Instead, freight must be consolidated from the depot to a satellite and then delivered from the satellite to the desired customer. This implicitly defines a two-echelon transportation system: the first level connecting the depot to the satellites and the second one, the satellites to the customers.

We define the depot with v_0, the set of satellites, with V_s, and the set of customers with V_c. Let n_s be the number of satellites, and n_c the number of customers. The customers are the destinations of the freight shipments and each customer i has an associated demand d_i representing the quantity of freight that has to be delivered to it. The demand of each customer cannot be split among different vehicles at the second level. For the first level, we consider that each satellite can be served by more than one first-level vehicle, therefore the aggregated freight assigned to each satellite can be split into two or more vehicles. Each first level vehicle can deliver the freight of one or several customers, as well as serve more than one satellite in the same route. We consider only one product, i.e., the volumes of freight belonging to different customers can be stored together and loaded into the same vehicle for both the first and the second-level movements. We define a route made up of a first-level vehicle starting from the depot, serving one or more satellites, and ending up at the depot, as *first-level route*. A *second-level route* is made up of a second-level vehicle starting from a satellite, serving one or more customers, and ending up at the same satellite. The fleet sizes are fixed and known in advance for both levels. All vehicles belonging to the same level have the same capacity. Each vehicle may perform at most one route. Satellites have limited capacity defined as the maximum number of vehicles that can leave from it. Different satellites may have different capacities.

The literature on *2E-VRP* is still somewhat limited. A general time-dependent formulation with fleet synchronization and customer time windows has been introduced in [2] in the context of two-echelon City Logistics systems. The authors have indicated promising algorithmic directions, but no implementation has been reported. A MIP formulation for the *2E-VRP* has been presented in [16], with instances with up to 32 customers solved to optimality. In the same paper, the authors derived two math-heuristics (MH) able to address instances with up to 50 customers. Both MH are based on the MIP model presented in the paper and work on the customer-to-satellite assignment variables. The first math-euristic, called *Diving*, considers a continuous relaxation of the model and applies a diving procedure to the customer-to-satellite assignment variables that are not integer. A restarting procedure is incorporated to recover possible unfeasibilities due to variable fixing. The second one is named *Semi-continuous*; in this method, the arc usage variables are considered continuous, while the assignment variables are still considered integer. The method solves this relaxed problem and uses the obtained values of the assignment variables to build a feasible solution for the *2E-VRP*. Several families of valid inequalities have been proposed in [15]. The valid inequalities are integrated into a branch-and-cut (BC) scheme, which is able to drastically reduce the optimality gap. A multi-start (MS) heuristic has been presented in [4]. The method is based on a clustering heuristic, which mainly works on the assignment between satellites and customers. The heuristic is used by the authors to solve large-sized instances with up to 250 customers. In [3] the authors study the effect of different spatial distributions on the total costs and a comparison with the standard CVRP solutions is given, while the impact of realistic situations in urban freight delivery where the travel costs are affected by components different from the distance, like fixed costs for using the arcs, operational costs, and environmental costs can be found in [5].

A problem from the literature quite similar to the *2E-VRP* is the truck and trailer routing problem (TTRP), in which the use of trailers (a commonly neglected feature in the VRP) is considered where customers are served by a truck pulling a trailer. However, due to practical constraints, including government regulations, limited manoeuvring space at customer sites, road conditions, etc., some customers may only be serviced by a truck. These constraints exist in many practical situations. This problem, as the *2E-VRP*, involves two routing levels strictly interconnected. The main difference with *2E-VRP* is that, while in the *2E-VRP* freight must pass through the satellites, because it must be delivered to the customer only by second level vehicles, in the TTRP the delivery of certain customers can be directly carried out by first level vehicles (truck and trailer) without passing through satellites. In [19], the single truck and trailer routing problem with satellite depots (STTRPSD), a particular version of the TTRP, is introduced. In STTRPSD a vehicle composed of a truck with a detachable trailer serves the demand of a set of customers reachable only by the truck without the trailer. This accessibility constraint implies the selection of locations to park the trailer before performing the trips to the customers. This version of the problem is the most similar to the *2E-VRP* while all deliveries must be carried out by the same kind of vehicle (truck without the trailer), even if, in this case, only one vehicle is considered, while in our problem several vehicles could be used to fulfil the customers demands.

3 GRASP with Path Relinking

GRASP is a multistart meta-heuristic for combinatorial optimization [9]. It consists of a constructive procedure based on a greedy randomized algorithm. In literature, this procedure is often combined with LS (see [6, 7, 8] for a detailed survey of the method and its applications). Path relinking is an intensification strategy that explores trajectories connecting high-quality solutions. Path relinking was suggested as an approach to integrate intensification and diversification strategies in the context of tabu search [10, 11] and then extended to other heuristic methods [18]. This approach generates new solutions by exploring trajectories that connect high-quality solutions by starting from one of these solutions, called the *starting* solution, and generating a path in the search space that leads towards the other solution, called *guiding* solution. Laguna and Martí adapted path relinking to the context of GRASP as a form of intensification [12]. The relinking in this context consists of finding a path between a solution found with GRASP and a chosen elite solution. Therefore, the relinking concept has a different interpretation within GRASP since the solutions found by two successive GRASP iterations are not linked by a sequence of moves. See [17] for a survey and numerous examples of GRASP with path relinking.

3.1 GRASP with Path Relinking for the 2E-VRP

The customer-to-satellite assignment problem plays a crucial role when addressing the *2E-VRP*. In fact, assuming one knows the optimal customer-to-satellite assignments, the *2E-VRP* can be partitioned into at most $n_s + 1$ CVRP instances, where

n_s is the number of satellites, one for the first-level and one for each satellite with at least one customer assigned to it. Thus, following the math-euristics presented in [16], and in the meta-heuristics in [4], we focus on the customer-to-satellite assignment by searching for the optimal assignment, delegating state-of-the-art CVRP methods to solve the corresponding subproblems. Both exact and heuristics methods from the literature are suitable to this purpose. Using exact methods we would obtain more precise results, while heuristics would require a more limited computational effort. After preliminary tests, we decided to use the hybrid meta-heuristic proposed in [14], which provides a good compromise between solution quality and computational time. Anyway, the effort required to evaluate the objective function, for a given assignment, is considerable. Then, heuristic methods involving large neighborhoods exploration are not suitable for the 2E-VRP, while procedures in which a rule, that allows to identify promising solutions, is applied are strongly preferable. In this work we propose a GRASP, which fits well with these requirements, combined with a path relinking strategy. Furthermore, intensification is applied only on promising GRASP solutions with a strong reduction of global computational time.

More precisely, The proposed method, GRASP-PR, consists of four main phases which will be described in detail:

1. A GRASP procedure;
2. A feasibility search (FS) phase to be applied if the solution is unfeasible;
3. A LS phase to improve a solution;
4. A path relinking phase.

The innovative aspect of this method is neither in its single components, which are well established in literature, nor in the meta-heuristic framework, but in the way the different components are combined within the framework. More in details, GRASP-PR works as follows. First of all an initial assignment is computed following the clustering constructive heuristic presented in [4] and the corresponding solution is kept as current best solution. At each iteration, a new assignment is built by means of the GRASP procedure, and the correspondent 2E-VRP solution is evaluated. If it is unfeasible, a repair procedure, named FS, is applied. If the solution is feasible and promising, i.e. it is better or within a threshold from the current best, an intensification phase made by a LS and a path-relinking heuristic is applied, otherwise it is discarded. A pseudocode of the algorithm is reported in Algorithm 1.

Note that the initial solution is computed in order to have a current feasible solution. It is considered as the current best solution at the first iteration of GRASP-PR and used to determine if the solutions obtained by the GRASP procedure are promising or not, i.e. if LS and path relinking should be applied or not. In the following subsections each component of the meta-heuristic framework is described in detail.

Algorithm 1 GRASP with Path Relinking

Compute an initial solution by means of the clustering constructive heuristic presented in [4];
while a maximum number of iterations $ITERMAX$ is not reached **do**
  Compute a solution *CS* by means of the GRASP procedure
  if *CS* is unfeasible **then**
    Try to repair *CS* by means of the Feasibility Search (FS)
    if *CS* is still unfeasible **then**
      Discard *CS*
  if *CS* is feasible or it has been successfully repaired by FS **then**
    if *CS* cost is within a given threshold t from the best solution cost **then**
      Apply the Local Search (LS)
      Apply Path Relinking
    else
      Discard *CS*
  else
    Discard *CS*

3.2 GRASP

The GRASP procedure assigns customers to satellites. The core of the GRASP procedure is the clustering constructive heuristic presented in [4], where the customers are assigned according to a less-distance-based rule. The assignment of customer i to satellite l is made with probability p_{il}

$$p_{il} = \frac{1 - \frac{D_{il}}{\sum_{l \in V_s} D_{il}}}{n_s - 1}, \tag{1}$$

where D_{ij} is the distance between customer i and satellite j. The rationale is to assign customer i to satellite j with a probability inversely proportional to the distance between them. The k assignments with the highest probability are considered and one of them is randomly selected.

The algorithm is adaptive. Thus, if by assigning customer i to satellite j the satellite capacity is exceeded, i.e., the number of vehicles required at the satellite is larger than the number of vehicles available for that satellite, or the global required number of vehicles is larger than the fleet size, the correspondent assignment becomes forbidden by setting its probability to zero. In this way, at each step of the greedy algorithm, previous choices are taken into account. When all customers are assigned to satellites, the original problem can be split into several CVRP subproblems, which are solved by means of the hybrid meta-heuristic in [14]. This procedure differs from the greedy algorithm used in [4] to find an initial solution because, in GRASP, the customer-to-satellite assignment is probabilistic, while in the greedy algorithm it follows a deterministic rule assigning each customer to its nearest available satellite. Notice that, the GRASP procedure does not include a LS phase to improve the routes, as this optimization is delegated to the meta-heuristic used to solve the CVRP subproblems.

3.3 Feasibility Search

The GRASP procedure does not guarantee the feasibility of the obtained solution, because, even when the satellite capacity is satisfied, the global fleet size constraint may be violated. When this happens, we try to rebuild a feasible solution by means of the FS procedure.

The FS does not imply a neighborhood exploration, it rather proceeds in a straightforward customer-moving procedure, aiming to empty vehicles that are in excess at some satellites. More in detail, customers are selected based on a distance criterion maximizing their distance from the satellite. A selected customer is then moved, from its assigned satellite (the one with the less filled vehicle) to another randomly chosen satellite, in order to free the exceeding vehicle. These moves are repeated until the global fleet size constraint is satisfied. If no move allows the feasibility of the obtained solution, this solution is discarded.

3.4 Local Search

The LS phase is performed only if the solution obtained by GRASP or GRASP and FS is both feasible and promising, i.e., its cost is better (or within a given threshold) than the cost of the best solution found so far. The procedure adopted for the LS is based on the Clustering Improvement algorithm presented in [1]. The neighborhood explored by the LS is composed by all the solutions differing by exactly one customer-to-satellite assignment from the ones of the current solution.

The order according to which the solutions in the neighborhood are analyzed is given by an assignment list. Customers are sorted in non decreasing order of the difference between the distance to the satellite to which they are assigned in the solution, and the distance to the nearest not-assigned satellite. The choice of this sorting order is based on the observation that customers displaying smaller such differences lead to improved solutions with a much higher frequency than the others. The exploration follows a First Improvement criterion and terminates after LS_{max} iterations or when the entire neighborhood has been explored without finding any improvement to the best solution. Preliminary computational experiments shown that this neighborhood exploration strategy is much more efficient than a standard random one, and that following a First Improvement criterion obtained better results than a Best Improvement one. A pseudocode of this procedure is reported in Algorithm 2.

Even if the neighborhood size is not so large, $O(n_c)$, where n_c is the number of customer, the computational time could grow up due to the need of recompute the CVRPs after a change in the customer-satellite assignments. This is the rationale of adding the additional heuristic stopping criterion when the reassignment has an objective function which is significantly worse than the current solution, i.e. it is larger of more than a given percentage threshold, γ, with respect to the current solution. In fact, being the customers ordered by non-decreasing order of RC_i and being

Algorithm 2 Local search

Given the current solution, the customers are sorted by non-decreasing order of the reassignment cost, defined as $RC_i = c_{ij} - c_{ik}$, where i is a customer, j is the satellite to which i is assigned in the current solution, and $k \neq j$ is the satellite such that, moving i from satellite j to satellite k, the capacity constraints on the global second-level vehicle fleet and the satellite k are satisfied and the cost c_{ik} is minimum among the satellites $k \neq j$. This is equivalent to order the customers according to non-decreasing order of the estimation of the change in the solution quality due to the assignment of one customer from the present satellite to its second-best choice. Let be CL the ordered list of the customers.

```
repeat
   Consider the first customer i in CL;
   if k exists then
      remove i from CL;
   else
      terminate the LS algorithm and return the best solution;
   Solve the CVRPs of satellites j and k;
   Update the demand of each satellite according to the new assignment and solve the first-level
   CVRP;
   Compute the objective function of the new solution and compare it to the cost of the current
   solution;
   if the new solution is better then
      Keep it as new current solution and exit from the neighborhood;
   else
      if the new solution has an objective function which is worse than a fixed percentage thresh-
      old γ from the objective function of the current solution then
         Terminate the LS algorithm and return the best solution;
      else
         Consider the next customer in the list
until CL is empty
```

RC_i related to the change in the objective function when we assign the customer to another satellite, if the objective function of a neighbor is deteriorating too much, it is unlikely that the following neighbors may bring us an improving solution.

3.5 Path Relinking

The path relinking phase consists of starting from the local optimum S_{LS} obtained by the LS procedure and "relinking" it to best solution S_b. The relinking is performed in a backward way, from S_b towards S_{LS}, inserting an element of S_{LS} into S_b at each step. More precisely, the path relinking procedure considers a customer assigned to satellite s_1 in S_{LS} and to satellite s_2 in S_b. It then assigns this customer to satellite s_2 in S_{LS}, without changing the other assignments. If the new solution is unfeasible, then the FS is applied. If it is still unfeasible it is discarded. The procedure terminates when S_{LS} becomes equal to S_b. The order according to which the customers are selected is given by a list in which customers are ordered in non decreasing order of the difference of distances between the customer and the satellites to which it is assigned in S_b and the customer and the satellites to which it has been assigned in S_{LS}. In this way we first analyze most promising moves, i.e., solutions characterized by a customer-satellite change minimizing its perturbation to the solution. Preliminary tests have shown the effectiveness of this path relinking strategy [13].

4 Computational Results

In this section, we present computational results and analyze the performance of the method we propose. An analysis of the impact of each component of the algorithm is reported and the results of GRASP-PR are compared with the literature.

Computational tests were effectuated on instances with 50 customers and 5 satellites, introduced in [3]. The instances present different combinations of customer distributions and types of satellite locations. Three customer distributions are considered representing a regional area, a large city, and a small town. Three types of satellite locations are considered as well, namely, random around the customer area, sliced around the customer area, and within part of the ring around the customer area, the latter representing city settings with limited accessibility due to geographical constraints (e.g., near to natural barriers such as the sea, a lake, or a mountain). Two instances were randomly generated for each combination of customer and satellite distributions. Table 1 displays the list of instances and their layout characteristics. Computational tests were performed on a computer with a Core 2 Duo processor at 2.5 GHz. The number of iterations, ITERMAX, and the maximum number of LS iterations, LS_{max}, were fixed to 25 and 250, respectively. This values come out of a tuning phase. In Table 2 we report best and average (over ten trials) results obtained by the different steps of the methods. More in details the table is organized as follows:

- Column 1: Instance name;
- Column 2: Results of GRASP and FS (phases 1 and 2 of the algorithm);
- Column 3: Results of GRASP, FS, and LS (phases 1, 2, and 3);
- Column 4: Results of the overall GRASP-PR, in which all phases are applied (GRASP, FS, LS, and path relinking);

A comparison with the literature is reported in Tables 3 and 4. More in detail, objective function values are reported in Table 3 while the corresponding computational times, expressed in CPU seconds, are reported in Table 4. Both tables are organized as follows:

- Column 1: Instance name;
- Column 2: Results of GRASP and FS (phases 1 and 2 of the algorithm);
- Column 3: Results of GRASP, FS, and LS (phases 1, 2, and 3);
- Column 4: Results of the overall GRASP-PR, in which all phases are applied (GRASP, FS, LS, and path relinking);

- Columns 5–7: Results of MS heuristic proposed in [4], the MH presented in [16], and the BC proposed in [15] with a time-limit of 10,000 s (BC also yields, on average, the overall best solutions in the literature).

Each row reports the values of a single instance, while the last two rows give the mean values and the gaps with respect to the BC (in Table 3 only), respectively.

When compared to the results of the MS heuristic (column MS), the pure GRASP obtains comparable results, while the introduction of LS yields a gain of more than 6%. The complete meta-heuristic GRASP-PR shows an improvement of 8.7% with respect to MS and outperforms the best heuristic results in the literature (MH) by 1.5%. This behavior confirms the trend reported in the literature, which encourages the use of path relinking as an enhancement of GRASP.

GRASP-PR achieves slightly worst results with respect to the exact solution method BC, but this small loss in accuracy (1.89%) is highly compensated by a significant reduction in computational effort of more than two orders of magnitude, as shown in Table 4.

Table 1: Instance layout characteristics

Instances	Customer distribution	Satellite location
Instance50-s5-37.dat	Regional area	Random
Instance50-s5-38.dat	Regional area	Random
Instance50-s5-39.dat	Regional area	Sliced
Instance50-s5-40.dat	Regional area	Sliced
Instance50-s5-41.dat	Regional area	Limited accessibility
Instance50-s5-42.dat	Regional area	Limited accessibility
Instance50-s5-43.dat	Large city	Random
Instance50-s5-44.dat	Large city	Random
Instance50-s5-45.dat	Large city	Sliced
Instance50-s5-46.dat	Large city	Sliced
Instance50-s5-47.dat	Large city	Limited accessibility
Instance50-s5-48.dat	Large city	Limited accessibility
Instance50-s5-49.dat	Small town	Random
Instance50-s5-50.dat	Small town	Random
Instance50-s5-51.dat	Small town	Sliced
Instance50-s5-52.dat	Small town	Sliced
Instance50-s5-53.dat	Small town	Limited accessibility
Instance50-s5-54.dat	Small town	Limited accessibility

5 Conclusion

We presented GRASP-PR, a GRASP with path relinking meta-heuristic for the *2E-VRP*, an extension of the classical vehicle routing problem, in which the delivery

Table 2: Best and average results

Inst	GRASP		GRASP+LS		GRASP-PR	
	Best	Avg	Best	Avg	Best	Avg
Instance50-s5-37.dat	1,599.86	1,615.34	1,586.23	1,586.23	1,545.99	1,545.99
Instance50-s5-38.dat	1,335.22	1,335.22	1,222.27	1,222.27	1,172.83	1,172.83
Instance50-s5-39.dat	1,657.27	1,657.27	1,580.19	1,580.19	1,535.28	1,535.28
Instance50-s5-40.dat	1,260.14	1,409.40	1,197.00	1,197.00	1,197.00	1,197.00
Instance50-s5-41.dat	1,817.17	1,817.17	1,687.96	1,687.96	1,687.96	1,687.96
Instance50-s5-42.dat	1,509.39	1,509.39	1,191.46	1,191.46	1,191.46	1,191.46
Instance50-s5-43.dat	1,607.28	1,607.99	1,603.56	1,603.56	1,593.06	1,593.06
Instance50-s5-44.dat	1,111.28	1,111.28	1,063.25	1,063.25	1,047.96	1,047.96
Instance50-s5-45.dat	1,801.99	1,801.99	1,480.32	1,480.32	1,480.32	1,480.32
Instance50-s5-46.dat	1,248.41	1,248.41	1,074.88	1,074.88	1,074.88	1,074.88
Instance50-s5-47.dat	1,807.40	1,807.40	1,786.17	1,786.17	1,683.13	1,683.13
Instance50-s5-48.dat	1,178.88	1,188.8	1,178.88	1,178.88	1,078.28	1,078.28
Instance50-s5-49.dat	1,697.96	1,705.9	1,539.89	1,546.77	1,500.39	1,510.98
Instance50-s5-50.dat	1,201.11	1,201.11	1,201.11	1,201.11	1,072.42	1,072.42
Instance50-s5-51.dat	1,590.00	1,590.00	1,535.18	1,535.18	1,435.83	1,435.83
Instance50-s5-52.dat	1,132.20	1,132.20	1,132.20	1,132.20	1,132.20	1,132.20
Instance50-s5-53.dat	1,599.09	1,599.09	1,598.66	1,598.66	1,598.66	1,598.66
Instance50-s5-54.dat	1,206.97	1,304.97	1,201.90	1,201.90	1,201.90	1,201.90
Mean	1,464.53	1,479.83	1,381.17	1,381.56	1,346.09	1,346.67

Table 3: Comparison of objective function values

Inst	GRASP	GRASP+LS	GRASP-PR	MS	MH	BC
Instance50-s5-37.dat	1,599.86	1,586.23	1,545.99	1,586.23	1,587.95	1,528.73
Instance50-s5-38.dat	1,335.22	1,222.27	1,172.83	1,340.49	1,186.02	1,187.39
Instance50-s5-39.dat	1,657.27	1,580.19	1,535.28	1,604.32	1,525.24	1,528.25
Instance50-s5-40.dat	1,260.14	1,197.00	1,197.00	1,387.28	1,226.79	1,179.64
Instance50-s5-41.dat	1,817.17	1,687.96	1,687.96	1,762.62	1,726.04	1,681.04
Instance50-s5-42.dat	1,509.39	1,191.46	1,191.46	1,559.39	1,324.38	1,232.87
Instance50-s5-43.dat	1,607.28	1,603.56	1,593.06	1,687.28	1,453.11	1,422.29
Instance50-s5-44.dat	1,111.28	1,063.25	1,047.96	1,227.26	1,063.64	1,061.25
Instance50-s5-45.dat	1,801.99	1,480.32	1,480.32	1,756.60	1,497.91	1,444.82
Instance50-s5-46.dat	1,248.41	1,074.88	1,074.88	1,148.31	1,173.12	1,068.50
Instance50-s5-47.dat	1,807.40	1,786.17	1,683.13	1,683.13	1,620.7	1,581.57
Instance50-s5-48.dat	1,178.88	1,178.88	1,078.28	1,319.96	1,122.18	1,092.32
Instance50-s5-49.dat	1,697.96	1,539.89	1,500.39	1,500.39	1,508.87	1,441.64
Instance50-s5-50.dat	1,201.11	1,201.11	1,072.42	1,131.65	1,170.89	1,089.67
Instance50-s5-51.dat	1,590.00	1,535.18	1,435.83	1,600.83	1,456.12	1,440.64
Instance50-s5-52.dat	1,132.20	1,132.20	1,132.20	1,145.54	1,185.05	1,109.52
Instance50-s5-53.dat	1,599.09	1,598.66	1,598.66	1,647.67	1,569.59	1,554.58
Instance50-s5-54.dat	1,206.97	1,201.90	1,201.90	1,201.90	1,189.14	1,135.39
Mean	1,464.53	1,381.17	1,346.09	1,460.60	1,365.93	1,321.12
Gap (%)	10.86	4.55	1.89	10.56	3.39	

Table 4: Comparison of computational times in CPU seconds

Inst	GRASP	GRASP+LS	GRASP-PR	MS	MH	BC
Instance50-s5-37.dat	5	25	33	2	71	10,000
Instance50-s5-38.dat	5	27	38	2	68	10,000
Instance50-s5-39.dat	5	30	37	2	66	10,000
Instance50-s5-40.dat	2	64	69	5	73	10,000
Instance50-s5-41.dat	7	55	59	32	97	10,000
Instance50-s5-42.dat	8	49	54	2	67	10,000
Instance50-s5-43.dat	7	36	48	1	66	10,000
Instance50-s5-44.dat	12	12	44	14	66	10,000
Instance50-s5-45.dat	3	13	13	1	73	10,000
Instance50-s5-46.dat	2	22	22	7	69	10,000
Instance50-s5-47.dat	28	80	90	53	76	10,000
Instance50-s5-48.dat	15	15	45	13	74	10,000
Instance50-s5-49.dat	4	8	34	21	86	10,000
Instance50-s5-50.dat	5	54	67	1	98	10,000
Instance50-s5-51.dat	5	53	65	1	82	10,000
Instance50-s5-52.dat	5	55	55	12	67	10,000
Instance50-s5-53.dat	5	38	38	1	45	10,000
Instance50-s5-54.dat	17	77	96	32	30	10,000
Mean	8	40	50	11	71	10,000

from a single depot to customers is managed by routing and consolidating the freight through intermediate facilities. Computational tests show that the method we propose outperforms the methods from the literature. Due to the good quality of the obtained solutions and the limited computational effort, GRASP-PR could be adopted both for long term planning and on-demand optimization.

References

1. Crainic, T.G., Mancini, S., Perboli, G., Tadei, R.: Clustering-based heuristics for the two-echelon capacitated vehicle routing problem. Tech. rep., CIRRELT 2008-46, CIRRELT, Montreal (2008)
2. Crainic, T.G., Ricciardi, N., Storchi, G.: Models for evaluating and planning city logistics systems. Transport. Sci. **43**, 432–454 (2009)
3. Crainic, T.G., Mancini, S., Perboli, G., Tadei, R.: Two-echelon vehicle routing problem: A satellite location analysis. Procedia Soc. Behav. Sci. **2**, 5944–5955 (2010)
4. Crainic, T.G., Mancini, S., Perboli, G., Tadei, R.: Multi-start heuristics for the two-echelon vehicle routing problem. Lect. Notes. Comput. Sc. **6622**, 179–190 (2011)
5. Crainic, T.G., Mancini, S., Perboli, G., Tadei, R.: Impact of generalized travel costs on satellite location in two-echelon vehicle routing problem. Procedia Soc. Behav. Sci. **39**, 195–204 (2012)
6. Feo, T.A., Resende, M.G.C.: Greedy randomized adaptive search procedures. J. Global Optim. **6**, 109–133 (1995)
7. Festa, P., Resende, M.G.C.: An annotated bibliography of GRASP - Part I: Algorithms. Int. T. Oper. Res. **16**, 1–24 (2009)
8. Festa, P., Resende, M.G.C.: An annotated bibliography of GRASP - Part II: Applications. Int. T. Oper. Res. **16**, 131–172 (2009)

9. Gendreau, M., Potvin, J.Y.: Handbook of Metaheuristics, 2nd edn. Springer, New York (2010)
10. Glover, F., Laguna, M.: Tabu Search. Kluwer, Boston (1997)
11. Glover, F., Laguna, M., Martí, R.: Fundamentals of scatter search and path relinking. Control Cybern. **29**(3), 653–684 (2000)
12. Laguna, M., Martí, R.: Grasp and path relinking for 2-layer straight line crossing minimization. INFORMS J. Comput. **11**, 44–52 (1999)
13. Mancini, S.: The two-echelon vehicle routing problem. Ph.D. thesis, Politecnico di Torino (2011)
14. Perboli, G., Pezzella, F., Tadei, R.: EVE-OPT: an hybrid algorithm for the capability vehicle routing problem. Math. Method. Oper. Res. **68**, 361–382 (2008)
15. Perboli, G., Tadei, R., Masoero, F.: New families of valid inequalities for the two-echelon vehicle routing problem. Electro. Notes. Discrete Math. **36**, 639–646 (2010)
16. Perboli, G., Tadei, R., Vigo, D.: The two-echelon capacitated vehicle routing problem: models and math-based heuristics. Transport. Sci. **45**, 364–380 (2011)
17. Resende, M.G.C., Ribeiro, C.C.: GRASP and path-relinking: Recent advances and applications. In: Proceedings of the 5th Metaheuristics International Conference (MIC 2003), kyoto (2003)
18. Resende, M.G.C., Ribeiro, C.C., Glover, F., Martí, R.: Scatter search and path-relinking: Fundamentals, advances, and applications. In: Gendreau, M., Potvin, J. (eds.) Handbook of Metaheuristics, pp. 87–107. Springer, New York (2010)
19. Villegas, J., Prins, C., Prodhon, C., Medaglia, A., Velasco, N.: GRASP/VND and multi-start evolutionary local search for the single truck and trailer routing problem with satellite depots. Eng. Appl. Artif. Intel. **23**, 780–794 (2010)

A Hybrid (1+1)-Evolutionary Strategy for the Open Vehicle Routing Problem

Andreas Reinholz and Holger Schneider

Abstract The Open Vehicle Routing Problem (OVRP) is a small but practically relevant modification of the well known Capacitated Vehicle Routing Problem, in which each route ends at the last served customer. We present how the OVRP and its constraints are modeled in the super customer framework, an integrated modeling and optimization framework for general vehicle routing problems. The framework provides several standard neighborhoods that are based on the path exchange operator and allows the use of accelerated delta function evaluations. With predefined templates, these neighborhoods can be used in mutation operators, evolutionary algorithms, neighborhood search, multiple neighborhood search and iterated variation selection procedures. For the OVRP, we used this framework to implement a hybrid combination out of a stochastic multiple neighborhood search and an (1+1)-evolutionary strategy. The performance of this hybrid (1+1)-evolutionary strategy was evaluated on 16 standard benchmark instances from literature. Best known solutions were computed for all 16 instances and new best solutions were found for seven instances.

1 A Modeling Concept for General Vehicle Routing Problems

Vehicle routing problems (VRPs) are hard optimization problems that are practically relevant in logistics (i.e. distribution networks to stores and shops or supply networks of production systems). From the modeling point of view, they can be

Andreas Reinholz
Institute of Air Transport and Airport Research, German Aerospace Center (DLR), Cologne, Germany
e-mail: Andreas.Reinholz@gmx.de

Holger Schneider
Department for Transportation Logistics, Fraunhofer Institute, Dortmund, Germany
e-mail: Holger.Schneider@tu-dortmund.de

L. Di Gaspero et al. (eds.), *Advances in Metaheuristics*, Operations Research/ Computer Science Interfaces Series 53, DOI 10.1007/978-1-4614-6322-1_8,

interpreted as an optimization problem on a directed graph. Each node represents a customer or depot location and arcs represent connections between them. For the coding of a solution (route plan) we use a vehicle oriented view and structure. Departing from its starting location, each vehicle processes the service demands of its customers using a sequence of adjacent arcs and nodes of the directed graph.

The optimization task is to find a set of valid routes with minimal overall costs. Each route is served by a vehicle that is restricted by problem specific constraints (e.g. load capacity limit, operation time limit) which must be considered in valid solutions. Routes are coded by a sequence of customer service demands (orders) that represents a directed path in the graph. Nodes and arcs are labeled with sets of resource variables and corresponding vehicle specific resource transformation functions. Therefore costs, demands and constraints can be modeled with resource sets and transformation functions.

Many constraints for VRPs such as capacity limits, pickup and delivery or time windows can be modeled in an elegant way with this concept. Please note that a similar concept of constraint handling with resources was independently developed by other researches [4, 9, 10]. For structural constraints (e.g. serve each customer exactly once, each route has to start at the depot) we used another concept that is based on the following idea: An initialization phase assures that only structurally valid solutions are produced and all variation operators (neighborhood generating operators) transform structurally valid solutions solely in structurally valid solutions. One major advantage of this concept is that the validity check of these constraints needs no CPU time during the search process. For further information about the Vehicle Routing Problem we recommend the work of Doerner et al. [5] and Toth et al. [23].

2 The Open Vehicle Routing Problem

The Open Vehicle Routing Problem (OVRP) is a small but practically relevant modification of the standard Capacitated Vehicle Routing Problem (CVRP), in which each route starts at the depot and ends at the last served customer. For example, pricing models for the use of external vehicles or outsourced transportation services lead to a cost structure where costs are only accounted from the depot until the last served customer. An instance of the OVRP is defined by following assumptions and requirements:

- A depot with location information (i.e. coordinates)
- A list of N customers with location information and demand information (size of the customers demand and the related service time at the customer)
- A distance matrix of size $(N+1) \times (N+1)$ with information about the shortest travel distance between all locations
- A sized fleet of homogeneous vehicles with limited capacity
- Structural constraints:
  - A vehicle route is a sequence of locations.

  - Each vehicle route has to start at the depot. Then, it solely moves straight forward to customer locations using the shortest distance.
  - A route plan is a set of vehicle routes, in which each customer has to be served exactly once.

- Capacity constraints:
  - The sum of the customer demands on each vehicle route does not exceed the capacity limit.
  - Every demand of a customer has to be less or equal the capacity limit of the used vehicle type.

- Route length or operation time constraints:
  - The route length of a vehicle route is defined as the sum of the travel distance between adjacent locations of the route.
  - The operation time of a vehicle route is defined as the sum of route length and service time at the served customers.
  - The operation time of each vehicle does not exceed the operation time limit.

- Objective function:
  - The overall objective is finding a valid route plan so that the overall route length is minimal.

This description of the OVRP is motivated from a logistics point of view. A mathematical or comprehensive introduction of the OVRP can be found in [7, 11].

3 Modeling the OVRP in the General Modeling Concept

Handling the structural constraints of the OVRP, each route is modeled with a starting depot node and a special virtual ending node. All virtual ending nodes have the specific feature that no resource consumption is induced by them or their incoming edges. The validation of route restrictions is modeled by two resources. In the following notation the function $Demand(i)$ indicates the demand of customer i, the function $DropTime(i)$ gives the drop time of customer i and the function $DrivingTime(i, j)$ gives the time of traveling between the nodes i and j. The *FreeSpace* resource checks the capacity restriction. At the starting depot it is initialized with the vehicle capacity limit:

$$FreeSpace := VehicleTypeCapacity$$

For each order in the route, the *FreeSpace* resource is transformed by reducing it with the demand of the order:

$$FreeSpace := FreeSpace - Demand(i)$$

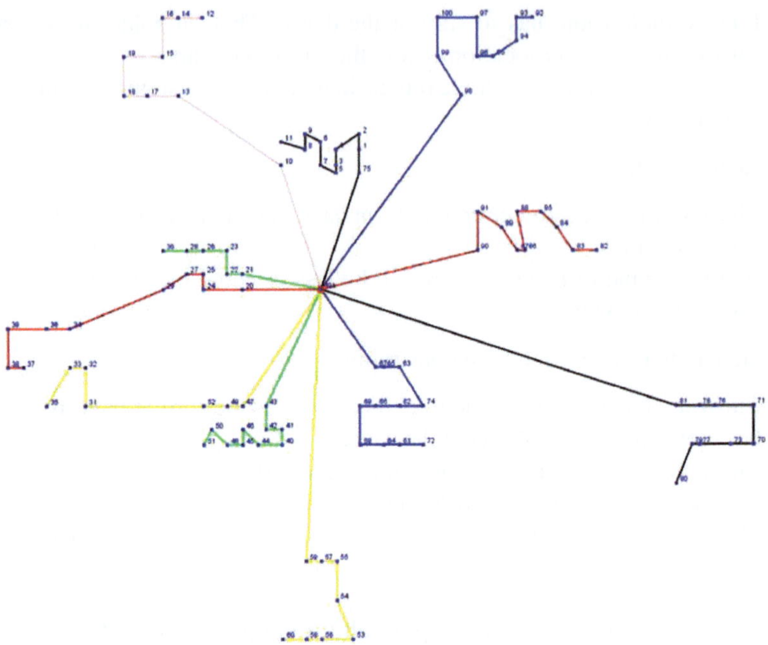

Fig. 1: OVRP solution to C14

For route length restriction we use the resource *RemainingRouteLength*. It is initialized with the maximal allowed route length:

$$RemainingRouteLength := MaxRouteLength$$

The transformation function of this resource distinguishes between nodes and edges:

$$RemainingRouteLength := RemainingRouteLength - DrivingTime(i,j) - DropTime(i)$$

For the objective function we used the standard one that minimizes the overall length of a route plan. The route length is computed by using the resource *RouteLength*:

$$RouteLength := RouteLength + DrivingTime(i,j)$$

Figure 1 illustrates the structure of OVRP solution with the best known solution to the 100 customer benchmark instance C14 from Christofides et al. [2].

4 Hybrid Metaheuristic

Our optimization approach uses the concept of iterative variation and selection (IVS) by Reinholz [16] describing a general framework for designing metaheuristic procedures. Complex solvers for heterogeneous optimization problems can be composed by combining more basic solvers in an iterative and recursive scheme. Similar ideas are proposed, for example, by Taillard et al. by adaptive memory programming (AMP) [22]. For this work, we designed a hybrid (1+1)-evolutionary strategy with the IVS concept that integrates a stochastic multiple neighborhood search in a (1+1)-evolutionary strategy.

4.1 Multiple Neighborhood Search

Our multiple neighborhood search generating template was previously described by Reinholz and Schneider [13, 14, 15, 18] and is based on the execution of different neighborhoods in a random order.

The neighborhoods are selected by a probability distribution that can be used to give different neighborhoods different priorities. These probability distribution and their parameters can be useful to utilize problem specific knowledge (e.g. generated in former runs). They can also be used as a link to incorporate adaptive or self-adaptive strategies during the search. In this work, we used the default uniform distribution that gives no special priorities to different neighborhoods.

In contrast to the concept of variable neighborhood search (VNS) and variable neighborhood descent (VND) of Hansen and Mladenovic [8], multiple neighborhood search [21] does not use a predefined deterministic sequence. The scheme of the stochastic multiple neighborhood search (MNS) template is given in Algorithm 1. A concrete multiple neighborhood search is determined by feeding the MNS template with a set of neighborhood searches through the pool variable NS_{ALL} with $\{NS_1, \ldots, NS_k\}$.

At the beginning of the multiple neighborhood search, the current neighborhood search working set NS_{SET} is initialized with all used neighborhood searches (NS_{ALL}). The process of choosing and executing selected neighborhood searches is iterated by a while loop. In each execution of the loop a neighborhood search is chosen randomly from NS_{SET} and is applied to the current best solution s_{best}. If a better solution than s_{best} is found by the selected neighborhood search, s_{best} is updated with the improving solution and NS_{SET} is refilled with all neighborhood searches from NS_{ALL} except the current selected neighborhood search. When no improvement was found, the last executed neighborhood search is removed from NS_{SET}. In both cases, the (updated) solution s_{best} is proven to be a local optimum with respect to the current selected neighborhood. The removal of a selected neighborhood search from the neighborhood search working set NS_{SET} avoids unnecessary repetitions of a pure random selection. The loop will be repeated until NS_{SET} is empty. This compound operator describes a probabilistic combination of the given

Algorithm 1 MNS(s_{init}, NS_{ALL}) % Multiple Neighborhood Search

Input: initial solution s_{init} and neighborhood set $NS_{ALL} := \{NS_1, \ldots, NS_k\}$

$s_{best} := s_{init}$
$NS_{SET} := NS_{ALL}$
while NS_{SET} is not empty **do**
    $NS := ChooseRandomNeighborhood(NS_{SET})$
    $s^* := NS(s_{best})$
    if s^* better than s_{best} **then**
        $s_{best} := s^*$
        $NS_{SET} := NS_{ALL}\ NS$
    else
        $NS_{SET} := NS_{SET}\ NS$

neighborhood searches. As long as improvements are found, NS_{SET} refills itself again and again. The loop ends, if none of the available neighborhood searches in NS_{ALL} can improve over s_{best}. Then, s_{best} is local optimal for all neighborhoods NS_{ALL}.

For vehicle routing problems and, in particular, for the OVRP, we use the so called path exchange operator as a basic neighborhood generating operator. The path exchange operator is motivated by an intuitive view of manipulating a route plan by systematically rearranging adjacent, non-overlapping service sequences (sub routes or paths) of a solution. The path exchange removes two non-overlapping paths by deleting four edges (4-opt) and reinserts each path at the former position of the other path. There are 4 different possibilities of reinserting the paths, because the inversion of the node sequence is allowed for each of the both paths.

Please note that one application of the path exchange operator affects only one (intra route exchange) or two routes (inter route exchange). Furthermore, the following neighborhood generating operators are special and restricted cases of the path exchange operator. They can be directly derived from the general path exchange operator by selecting restricted paths in the following way:

- Path move: exchange a nonempty path with an empty path
- 2-Opt: move and reverse a path to the same position
- Relocate: move a path of size 1
- Swap: exchange two paths of size 1
- End restricted path exchange: cut both routes somewhere into two paths and exchange the end pieces
- Route melting: move a path consisting of all customers of one route to the end or beginning of another route
- Route splitting: move a path to an empty route

Integrated in a neighborhood search template, each of these restricted path exchange operators induces a neighborhood that is also a subset of the neighborhood of the path exchange operator.

Algorithm 2 Mutation(s) % Mutation

Input: initial solution s
  $k := 1 + Binomial\_Distribution(MutationRate)$
  for $i := 1$ to k **do**
    $s := Mutate(s)$
  return s

Algorithm 3 Hybrid (1+1)-ES($s_{init}, \{NS_1, \ldots, NS_k\}, s_{best}$)

Input: initial solution s_{init} and neighborhood set $NS_{ALL} := \{NS_1, \ldots, NS_k\}$
  $s_{best} := s_{init}$
  $NS_{SET} := NS_{ALL}$
  while *termination criterion not met* **do**
    $s' := Mutation(s_{best})$
    $s_l := MNS(NS_{ALL}, s')$
    $s_{best} := Update(s_{best}, s_l)$
  return s

4.2 Hybrid (1+1)-Evolutionary Strategy

Our hybrid (1+1)-evolutionary strategy is a hierarchical composition of a standard (1+1)-evolutionary strategy and a stochastic multiple neighborhood search.

The (1+1)-ES is a well-known and well-analyzed evolutionary strategy (ES) that organizes an iterative search process starting from a single solution (first parent solution). In each iteration (generation), a modified child solution is generated by applying a probabilistic copy-and-variation scheme (mutation operator) to the actual parent solution. Then, an elitist selection scheme is applied to the parent and child solution (survival of the fittest) to determine the surviving solution that is taken as the parent solution of the next generation. This generation loop is iterated until a predefined termination criterion aborts the search process.

To create a hybrid (1+1)-evolutionary strategy, our stochastic MNS is inserted into the (1+1)-ES by placing it directly after the mutation operator. Many research studies have shown that the (1+1)-ES is a simple but powerful standard solver for many kinds of optimization problems. One advantage is the small number of parameters controlling the search process. Combined with neighborhood search procedures to a hybrid (1+1)-evolutionary strategy, it could be also interpreted as an iterated local search using the mutation operator as a probabilistic kick move. Using the neighborhood search metaphor, the mutation operator itself can be seen as a probabilistic choice of a neighbor in the k-step neighborhood of the current best solution s_{local}. The step size k is chosen in each iteration using a binomial distribution with parameter *MutationRate*, where $k = 0$ is avoided by construction. Algorithms 2 and 3 show the pseudo code for these procedures.

5 Acceleration Methods

In this section we outline two powerful acceleration techniques for neighborhood based search methods that reutilize partial information from already visited search points and neighborhoods.

5.1 Accelerated Evaluation of the Objective Function

Neighborhood generating operators may have the property that only small parts of the candidate solutions are modified. Methods for *accelerated delta function evaluation* try to utilize this to speed up the computation of the objective function and the constraints check by reusing intermediate results. The task is to find a clever way of organizing and storing intermediate (partial) evaluations for the considered combination of objective functions, constraints and neighborhood generating operators.

Our acceleration techniques for the OVRP and the path exchange operator are based on the *Super Customer Concept* of Reinholz [14, 15, 17]. The idea is to handle each path like a single super customer and to save objective functions and constraints related information of all possible sub paths of a candidate solution in a super customer memory. This memory is pre-computed at the beginning of a neighborhood search. For many constraints (e.g. capacity limit, route length limit, time windows, heterogeneous fleet, etc.) the super customer concept allows the use of an accelerated delta function evaluation for path exchange operators that evaluates the objective function and the constraints in constant time: The evaluation of a path exchange move reduces to the concatenation of at most six super customers (see Fig. 2). For more information see also [16, 17, 18].

5.2 Decomposed and Accelerated Neighborhood Search

The second acceleration procedure uses another decomposition method to speed up the neighborhood search process itself. For this we benefit from the local property of VRP where routes are nearly independent entities. When using the path exchange operator as an elementary neighborhood generating operator, each neighboring candidate solution has at most two modified routes r_i and r_j. Therefore the neighborhood search for an initial solution S can be decomposed in a set of neighborhood searches of all possible route pairs. So the task of finding an improving solution in neighborhood is decomposed into a set of independent sub-tasks or sub-problems for each route pair. Figure 3 shows the general idea of reduced evaluation effort by observing overlapping neighborhoods during iterative search process.

When selecting an (improving) neighbor solution S' of S for the next iteration of the neighborhood search, we can benefit from the fact that both route plans S' and S have many identical *route pair sub-problems*: *route pair sub-problems* that have

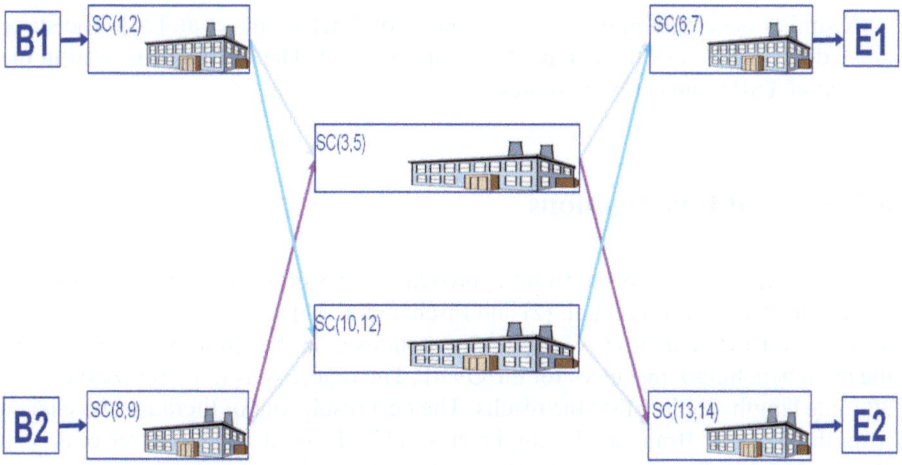

Fig. 2: Path exchange of super customers SC3-5 and SC10-12

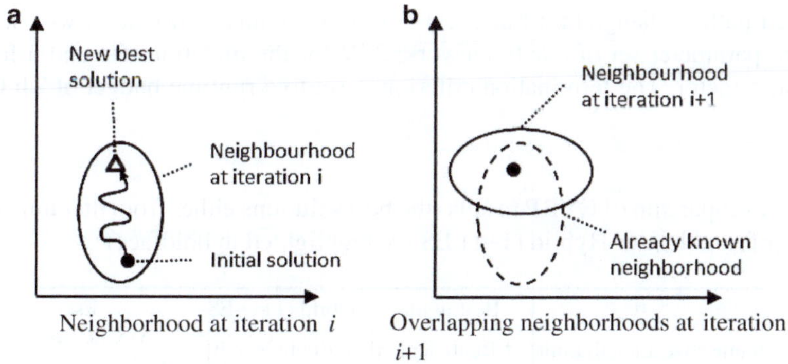

Fig. 3: Visual presentation of overlapping neighborhoods during iterative search process (**a**) Neighborhood at iteration i (**b**) Overlapping neighborhoods at iteration $i+1$

been solved already for S (or former candidate solutions). These do not need to be solved again for S'.

The neighborhood of a route plan S with k routes can be computed by solving $O(k^2)$ *route pair sub-problems*. To store the results of the *route pair sub-problems* we used a memory that is implemented as a matrix in which each cell (i, j) contains the result of the *route pair sub-problem* of route i and j. When proceeding from a route plan S to a neighboring route plan S' by changing route i and j, only the two columns and two rows with number i and j of the matrix have to be updated. Therefore, the set of *route pair sub-problems* of S' only contains at most $O(k)$ sub-problems which might not be in the set of *route pair sub-problems* of S.

A similar acceleration method is proposed by Zachariadis et al. [24]. The static move descriptor stores the potential for improvement. The technique is used in the context of TSP routes or giant routes.

6 Empirical Investigations

The capability of our optimization approach was tested on the 16 benchmark instances from Christofides et al. [2] and Fischer et al. [6]. Although being originally designed for the standard CVRP, the Christofides et al. benchmarks are nowadays the main benchmark instances for the OVRP. The objective is to minimize the overall route length. Table 1 lists the results. The best results out of literature were taken from the work of Brandao [1], Ropke et al. [12], Li et al. [11], Fleszar et al. [7], Derigs et al. [3], Repoussis et al. [19], Zachariadis et al. [24] and Salari et al. [20].

Our hybrid (1+1) evolutionary strategy used a multiple neighborhood search with the four standard neighborhood generating operators 2-Opt, relocate, swap and end restricted path exchange (see Sect. 4.1). For all benchmark instances, we used the standard parameter set of our framework: $2/N$ for the mutation rate and 5 for the multi-start factor. The termination criterion is set to a runtime budget of 2 h CPU-time.

Table 1: Comparison of OVRP results (the best solutions either from literature (Best in lit.) or found by our Hybrid (1+1)-ES are highlighted in boldface)

			Best in lit.		Hybrid (1+1)-ES			Est.
Name	Size	Length limit	#	Route length	#	Route length	Δ [%]	CPU sec.
C01	50	–	6	**412.95**	6	**412.95**	0.00	5.43
C02	75	–	11	**564.06**	11	**564.06**	0.00	16.74
C03	100	–	9	**639.25**	9	**639.25**	0.00	1,123.09
C04	150	–	12	**733.13**	12	**733.13**	0.00	602.68
C05	200	–	17	868.88	17	**868.44**	0.00	5,460.71
C06	50	200	6	**412.95**	6	**412.95**	0.00	3.63
C07	75	160	11	568.49	11	**566.93**	−0.27	53.02
C08	100	230	9	644.63	9	**642.11**	−0.003	452.79
C09	150	200	13	756.14	14	**741.44**	−1.94	1,196.68
C10	200	200	17	874.71	18	**871.58**	−0.36	5,482.56
C11	120	–	10	**678.54**	10	**678.54**	0.00	83.55
C12	100	–	10	**534.24**	10	**534.24**	0.00	5.51
C13	120	720	11	894.19	11	**836.55**	−6.45	2,463.99
C14	100	1040	11	581.81	11	**552.64**	−5.01	29.95
F11	72	–	4	**177**	4	**177**	0.00	391.77
F12	135	–	7	**769.55**	7	**769.55**	0.00	2,121.21
						Avg	−0.87	
						Max	−6.45	

Table 1 gives the objective function value of the best result out of five runs. The time for the first occurrence of the best found solution ranged from 3.63 s for C06 to 5,482.56 s for C10.

The comparison with the best results from the literature shows that the presented hybrid (1+1)-evolutionary strategy was capable to find best known results for all instances. For seven instances we could improve the best results from the literature between 0.003% and 6.45%. The average improvement over all instances is 0.87%.

Please note that the implementation of our VRP framework was done in JAVA and that the implemented constraints and modeling concept, which was originally designed to handle Rich Vehicle Routing Problems with many constraints and modeling aspects, generates a lot of unnecessary overhead when being applied to a low constraint VRP variant such as the OVRP. Comparisons with former work for other VRP variants, which were implemented in C++, indicate that speed up of 10 is here entirely realistic. The experiments were done on a standard notebook with an Intel Core 2 Duo with 2.4 GHz using only one core per run.

All results were found by using the same parameter setting for all benchmark instances. The standard parameter setting of the framework has also been used successfully for a wide range of other VRP variants from both literature and the real world. The intention of using the standard parameter setting was to show that the integrated modeling and optimization concept behind the framework can consistently produce high quality results without using a computational expensive parameter tuning or problem specific features.

Especially the strong improvement of the best known solution for instance C14 shows that the benchmarks from Christofides et al. [2] are still challenging for modern metaheuristics in the context of the OVRP. Tables 2 and 3 in the appendix present the details and features of the new best solution for instance C14.

7 Conclusion

We have presented how the Open Vehicle Routing Problem can be elegantly modeled with our integrated modeling and optimization approach for rich vehicle routing problems. Using predefined templates, we implemented a hybrid metaheuristics in our framework that uses advanced concepts out of the research fields multiple neighborhood search, evolutionary algorithms and neighborhood search acceleration techniques.

The presented framework was designed to handle general rich vehicle routing problems with all kinds of complex constraints and objectives that arise in the real world. In the context of real world problems, a flexible modeling and optimization instrument with the capability to follow vividly and changing optimization tasks (rapid prototyping) is often more crucial than having a high performing but specialized solver. Nevertheless, on a set of 16 standard benchmark instances of the Open Vehicle Routing Problem we could show that our general approach produces consistently high quality solutions: For all 16 instances we found the best known solutions and for 7 instances we were able to produce new best results.

The intention of this paper is to show that such a simple but general approach can also produce high quality results for benchmarks of a specific class of vehicle routing problems without using problem specific features or parameter tuning.

Appendix: Best Known Solution for C14

Tables 2 and 3 present the details and features of the new best solution for instance C14. For each of the 11 routes there is a block of six rows that helps to validate the solution:

1. The sequence of the customer IDs. The Depot is indicated by a 0
2. The distance between two adjacent customers in row 1
3. The summed up route length until the customers in row 1
4. The summed up route length and drop time (operation time)
5. The demand of the customer in row 1. The demand for the depot is 0
6. The summed up route demand until the customers in row 1

Table 2: Solution details for OVRP instance C14 (part 1)

Route id	Type	Stops										
1	Stops	0	81	78	76	71	70	73	77	79	80	
	Distance	47.4	3.0	2.0	5.0	5.0	3.0	4.0	1.0	5.3		
	$\sum$ length	47.4	50.4	52.4	57.4	62.4	65.4	69.4	70.4	75.7	75.7	
	$\sum$ OpTime	47.4	140.4	232.4	327.4	422.4	515.4	609.4	700.4	795.7	885.7	
	Demand	0	30	20	10	20	30	10	10	10	10	
	$\sum$ demand	0	30	50	60	80	110	120	130	140	150	
2	Stops	0	20	24	25	27	29	34	36	39	38	37
	Distance	10.0	5.0	2.0	2.0	3.6	13.0	3.0	5.0	5.0	2.0	
	$\sum$ length	10.0	15.0	17.0	19.0	22.6	35.6	38.6	43.6	48.6	50.6	50.6
	$\sum$ OpTime	10.0	105.0	197.0	289.0	382.6	485.6	578.6	673.6	768.6	860.6	950.6
	Demand	0	10	10	40	10	10	20	10	20	30	20
	$\sum$ demand	0	10	20	60	70	80	100	110	130	160	180

Table 3: Solution details for OVRP instance C14 (continued)

Route id	Type	Stops											
3	Stops	0	98	99	100	97	96	95	94	93	92		
	Distance	30.8	5.8	5.0	5.0	5.0	2.0	3.6	3.0	2.0			
	$\sum$ length	30.8	36.6	41.6	46.6	51.6	53.6	57.2	60.2	62.2	62.2		
	$\sum$ OpTime	30.8	126.6	221.6	316.6	411.6	503.6	597.2	690.2	782.2	872.2		
	Demand	0	20	10	20	30	10	30	10	40	20		
	$\sum$ demand	0	20	30	50	80	90	120	130	170	190		
4	Stops	0	21	22	23	26	28	30					
	Distance	10.1	2.0	3.0	3.0	2.0	3.0						
	$\sum$ length	10.1	12.1	15.1	18.1	20.1	23.1	23.1					
	$\sum$ OpTime	10.1	102.1	195.1	288.1	380.1	473.1	563.1					
	Demand	0	20	20	10	10	20	10					
	$\sum$ demand	0	20	40	50	60	80	90					
5	Stops	0	59	57	55	54	53	56	58	60			
	Distance	35.0	2.0	2.0	5.0	5.3	4.0	2.0	3.0				
	$\sum$ length	35.0	37.0	39.0	44.0	49.3	53.3	55.3	58.3	58.3			
	$\sum$ OpTime	35.0	127.0	219.0	314.0	409.3	503.3	595.3	688.3	778.3			
	Demand	0	10	40	10	40	20	30	30	20			
	$\sum$ demand	0	10	50	60	100	120	150	180	200			
6	Stops	0	10	13	17	18	19	15	16	14	12		
	Distance	16.7	15.8	4.0	3.0	5.0	5.0	5.0	2.0	3.0			
	$\sum$ length	16.7	32.5	36.5	39.5	44.5	49.5	54.5	56.5	59.5	59.5		
	$\sum$ OpTime	16.7	122.5	216.5	309.5	404.5	499.5	594.5	686.5	779.5	869.5		
	Demand	0	10	30	20	20	10	40	40	10	20		
	$\sum$ demand	0	10	40	60	80	90	130	170	180	200		
7	Stops	0	75	1	2	4	3	5	7	6	9	8	11
	Distance	15.8	3.0	2.0	3.6	2.0	1.0	2.2	3.0	2.2	2.0	3.1	
	$\sum$ length	15.8	18.8	20.8	24.4	26.4	27.4	29.6	32.6	34.8	36.8	39.9	39.9
	$\sum$ OpTime	15.8	108.8	200.8	294.4	386.4	477.4	569.6	662.6	754.8	846.8	939.9	1029.9
	Demand	0	20	10	30	10	10	10	20	20	10	20	10
	$\sum$ demand	0	20	30	60	70	80	90	110	130	140	160	
8	Stops	0	90	91	89	87	86	88	85	84	83	82	
	Distance	20.6	5.0	3.6	3.6	1.0	5.0	3.0	2.8	3.6	3.0		
	$\sum$ length	20.6	25.6	29.2	32.8	33.8	38.8	41.8	44.6	48.2	51.2	51.2	
	$\sum$ OpTime	20.6	115.6	209.2	302.8	393.8	488.8	581.8	674.6	768.2	861.2	951.2	
	Demand	0	10	10	10	20	10	30	30	20	10	20	
	$\sum$ demand	0	10	20	30	50	60	90	120	140	150	170	
9	Stops	0	67	65	63	74	62	66	69	68	64	61	72
	Distance	12.2	1.0	2.0	5.8	3.0	3.0	2.0	5.0	3.0	2.0	3.0	
	$\sum$ length	12.2	13.2	15.2	21.0	24.0	27.0	29.0	34.0	37.0	39.0	42.0	42.0
	$\sum$ OpTime	12.2	103.2	195.2	291.0	384.0	477.0	569.0	664.0	757.0	849.0	942.0	1032.0
	Demand	0	10	10	50	50	20	10	10	10	10	10	10
	$\sum$ demand	0	10	20	70	120	140	150	160	170	180	190	200
10	Stops	0	43	42	41	40	44	46	45	48	50	51	
	Distance	16.5	3.0	2.0	2.0	3.0	2.8	2.0	2.0	2.8	2.2		
	$\sum$ length	16.5	19.5	21.5	23.5	26.5	29.3	31.3	33.3	36.1	38.3	38.3	
	$\sum$ OpTime	16.5	109.5	201.5	293.5	386.5	479.3	571.3	663.3	756.1	848.3	938.3	
	Demand	0	10	20	10	10	10	30	10	10	10	10	
	$\sum$ demand	0	10	30	40	50	60	90	100	110	120	130	
11	Stops	0	47	49	52	31	32	33	35				
	Distance	18.0	2.0	3.0	15.0	5.0	2.0	5.8					
	$\sum$ length	18.0	20.0	23.0	38.0	43.0	45.0	50.8	50.8				
	$\sum$ OpTime	18.0	110.0	203.0	308.0	403.0	495.0	590.8	680.8				
	Demand	0	10	10	10	20	30	40	10				
	$\sum$ demand	0	10	20	30	50	80	120	130				

References

1. Brando, J.: A tabu search algorithm for the open vehicle routing problem. Eur. J. Oper. Res. **157**(3), 552–564 (2004)
2. Christofides, N., Mingozzi, A., Toth, P.: Combinatorial optimization. The Vehicle Routing Problem, pp. 315–338. Wiley, Chichester (1979)
3. Derigs, U., Reuter, K.: A simple and efficient tabu search heuristic for solving the open vehicle routing problem. J. Oper. Res. Soc. **60**(12), 1658–1669 (2009)
4. Desaulniers, G., Desrosiers, J., Ioachim, I., Solomon, M.M., Soumis, F., Villeneuve, D.: A unified framework for deterministic time constrained vehicle routing and crew scheduling problems. Fleet Management and Logistics, pp. 57–93. Kluwer, Dordrecht (1998)
5. Doerner, K., Schmid, V.: Survey: matheuristics for rich vehicle routing problems. In: Blesa, M., Blum, C., Raidl, G., Roli, A., Sampels, M. (eds.) Hybrid Metaheuristics. Lecture Notes in Computer Science, vol. 6373, pp. 206–221. Springer, Berlin (2010)
6. Fisher, M., Jakumar, R.: A generalized assignment heuristic for vehicle routing. Networks **11**, 109–124 (1981)
7. Fleszar, K., Osman, I.H., Hindi, K.S.: A variable neighbourhood search algorithm for the open vehicle routing problem. Eur. J. Oper. Res. **195**(3), 803–809 (2009)
8. Hansen, P., Mladenovic, N.: Variable neighborhood search: principles and applications. GERAD and cole des Hautes tudes Commerciales, Montral, Canada (1997)
9. Irnich, S.: Resource extension functions: properties, inversion an generalization to segments. Technical report, Deutsche Post Endowed Chair of Opimization of Distribution Networks, RWTH Aachen, Universitt, Aachen, Germany (2006)
10. Kindervater, G., Savelsberg, M.: Vehicle routing: handling edge exchanges. Local search in combinatorial optimization, pp. 337–360. Wiley, Chichester (1997)
11. Li, F., Golden, B., Wasil, E.: The open vehicle routing problem: algorithms, large-scale test problems, and computational results. Comput. Oper. Res. **34**(10), 2918–2930 (2007)
12. Pisinger, D., Ropke, S.: A general heuristic for vehicle routing problems. Technical report 8, DIKU, Department of Computer Science, University of Copenhagen, Denmark (2007)
13. Reinholz, A.: Genetische Algorithmen: Transportoptimierung und Tourenplanung für ein zentrales Auslieferungsdepot. Der GMD-Spiegel **2**, 20–24 (1995)
14. Reinholz, A.: Transportoptimierung und Tourenplanung fr ein zentrales Auslieferungsdepot. Intermediate Report of the Project HYBRID, GMD, St. Augustin, Germany (1995)
15. Reinholz, A.: How to design efficient iterated local search, variable neighborhood search and hybrid evolutionary algorithms for periodic multi depot vehicle routing problems (tutorium). In: 5th Metaheuristics International Conference (MIC 03), Kyoto, Japan (2003)
16. Reinholz, A.: Optimizing logistics networks with metaheuristics. In: Optimization of Logistics Systems, Methods and Experiences, pp. 85–98. Symposium of the Collaborative Research Center 559 "Modeling of Large Logistics Networks", Verlag Praxiswissen, Fraunhofer IML, Dortmund, DE (2008)
17. Reinholz, A., Manthey, G.: Umsetzung und Leistungsbewertung des Ressourcemengen- und Superkundenkonzepts für erweiterte Vehicle Routing Probleme. Technical report 08009, Collaborative Research Centre 559, Modeling of Large Logistics Networks (2008)
18. Reinholz, A., Schneider, H.: Ein Prozess- und Objektorientiertes Modellierungskonzept fr Rich VRP. Springer, Hamburg (2009)
19. Repoussis, P., Tarantilis, C., Brysy, O., Ioannou, G.: A hybrid evolution strategy for the open vehicle routing problem. Comput. Oper. Res. **37**(3), 443–455 (2010)
20. Salari, M., Toth, P., Tramontani, A.: An ilp improvement procedure for the open vehicle routing problem. Comput. Oper. Res. **37**(12), 2106–2120 (2010)
21. Sörensen, K., Sevaux, M., Schittekat, P.: Multiple neighborhood search in commercial VRP packages: evolving towards self-adaptive methods. Lecture Notes in Economics and Mathematical Systems, vol. 136. Springer, London (2008)
22. Taillard, E.D., Gambardella, L.M., Gendreau, M., Potvin, J.Y.: Adaptive memory programming: a unified view of metaheuristics. Eur. J. Oper. Res. **135**(1), 1–16 (2001)

23. Toth, P., Vigo, D.: An Overview of Vehicle Routing Problems, pp. 1–26. Society for Industrial and Applied Mathematics, Philadelphia (2001)
24. Zachariadis, E.E., Kiranoudis, C.T.: An open vehicle routing problem metaheuristic for examining wide solution neighborhoods. Comput. Oper. Res. **37**(4), 712–723 (2010)

A Timeslot-Filling Heuristic Approach to Construct High-School Timetables

Michael Pimmer and Günther R. Raidl

Abstract This work describes an approach for creating high-school timetables. To develop and test our algorithm, we used the international, real-world instances of the *Benchmarking project for (High) School Timetabling*. Contrary to most other heuristic approaches, we do not try to iteratively assign single meetings (events) to timeslots. Instead, we repeatedly choose a not entirely occupied timeslot and aim at simultaneously assigning the most suitable set of meetings. To improve and diversify the solutions, a heuristic that deletes and reassigns certain timeslots, events or resources is applied and combined with a hill-climbing procedure to find suitable parameters for grading constraints. Experimental results indicate the competitiveness of this new approach.

1 Introduction

The task of high-school timetabling is to assign events—normally class–teacher meetings—to rooms and timeslots of a weekly schedule. Dependent on the country and institution of origin, the requirements and specifics of the problem vary drastically. As no commonly accepted, international benchmark instances existed until recently, most scientists have been working with specific local instances or strongly simplified models as the classical Class-Teacher Timetabling Problem (CTTP) described by Gotlieb [8] in 1974. Thus, most of this work is hard to compare or of limited practical relevance.

In 2007, the Benchmarking Project for (High) School Timetabling was launched to settle this issue. Based on an XML file format and a well-defined evaluation function, it currently provides more than 20 real-world instances from various countries. The XML-format describes *resources* (e.g. teachers, students, rooms) which can be

Michael Pimmer · Günther R. Raidl
Institute of Computer Graphics and Algorithms, Vienna University of Technology, Vienna, Austria
e-mail: michael@pimmer.info, raidl@ads.tuwien.ac.at

L. Di Gaspero et al. (eds.), *Advances in Metaheuristics*, Operations Research/ Computer Science Interfaces Series 53, DOI 10.1007/978-1-4614-6322-1_9,

part of *events*. An event is a meeting that usually requires some resources and should be assigned to a timeslot. If any resource out of a set of resources—e.g. any English teacher—shall be assigned to an event, this is called *open role*. Further information about the project is available in [16] and on the project website.[1]

While most existing heuristics to solve school timetabling problems are based on an iterative process that assigns single events to timeslots, we follow here the concept of repeatedly choosing a not entirely occupied timeslot and assigning a promising larger set of suitable meetings at the same time. In addition an improvement procedure that deletes and reassigns timeslots, events, or resources is applied in combination with a hill-climbing procedure for adapting parameters controlling the grading of constraints. This new approach was specifically developed with the instances of the Benchmarking Project for (High) School Timetabling and general applicability in mind.

The next section gives an overview on related work. Our approach is described in Sects. 3 and 4 presents and discusses achieved results, and conclusions are drawn in Sect. 5.

2 Related Work

Concerning the problem complexity, the classical CTTP was shown to be NP-complete when any unavailabilities are given [7]. Relaxing the restrictive definition of meetings and adding some common constraints from real-world school timetabling problems introduces significant additional complexity. Kingston and Cooper [10] identified five NP-hard subproblems, and Willemen [20] extended these by two more.

A broad range of approaches has been applied to high-school timetabling. For extensive information on existing methods, the reader is referred to well-known surveys [3, 17, 14] and to the international conferences *Practice And Theory of Automated Timetabling (PATAT)*[2] [4, 2, 12] as well as to the EURO working group on automated timetabling *EURO-WATT*.[3]

The most common solving strategy is to iteratively assign single events, combined with a backtracking on dead ends. The less explored alternative we consider in this work is to create a timetable by repeatedly filling selected timeslots. This approach was first mentioned by Schmidt and Ströhlein [18] in 1972. Unfortunately, their formulation of a CTTP as a vertex coloring problem is not always applicable to real-world instances, because many constraints have to be considered additionally to the coloring-problem. For example, problems arise with events of longer duration requiring multiple timeslots, especially if it is not determined how an event can be split, or in case of multiple open roles requiring the same resource. Considering

[1] http://www.utwente.nl/ctit/hstt/.

[2] http://www.asap.cs.nott.ac.uk/patat/patat-index.shtml.

[3] http://www.asap.cs.nott.ac.uk/watt/.

the vertex coloring formulation, not only the weights and possible colors of edges would change during the solving procedure, but nodes and edges may even appear or disappear.

In 2002, Abraham and Kingston [1] applied the timeslot-filling approach to an Australian instance. They pre-calculate possible compatible sets of events. As there are too many sets, smaller events which are supposed to be easier to schedule are omitted. Then, a set is chosen for each timeslot in a way to cover all required events. Finally, the omitted events are scheduled. Instead of continuing this work, effort was put in the creation of the KTS High School Timetabling System [9], which yielded better results.

KTS can be considered a hybrid approach. Such approaches lie in between assigning single events and filling timeslots at once. The goal is to break down the problem into easier subproblems by grouping events, without losing too much flexibility later on. This can be done by pre-assigning events to days, e.g. as described in [6], before assigning them to specific timeslots. Kingston [9] breaks down the problem by grouping events to tiles. The tiles are scheduled separately and joined afterwards to obtain the final timetable. This approach yields good results for Australian real-world instances in short runtimes and can to some degree be extended to be applicable to instances of other countries.

Some very Large-Scale Neighborhood Search Techniques (VLNS) delete and re-assign a major part of the solution. They were already applied to timetabling problems, e.g. by Meyers and Orlin [13]. The refilling-strategies described in Sect. 3.3 also belong to this kind of search technique.

3 Timeslot-Filling Heuristic

The central building block of our algorithm is to fill a timeslot with a suitable set of events. Consequently, we call it *Timeslot-Filling Heuristic* (TFH). We iteratively fill non-full timeslots as shown in Fig. 1 and further detailed in the subsequent sections. The higher-level strategies based on this building block are described in Sect. 3.3.

Because of the multitude of constraints considered in the benchmark project's instances and thus in our algorithm, we cannot present all details here but refer to the first author's master thesis [15] for an exhaustive description.

3.1 Grading of Events

The grade of an event represents the favorability of holding it in a chosen timeslot. It is defined as the sum of the grades of the constraints that apply to this event, that apply to any resource it requires (or that can fill any of its open roles), and that apply to any event group this event belongs to. Moreover, we calculate and add grades that are not directly related to constraints, but instead aim at completely filling the timetable (*timetable-filling grades*).

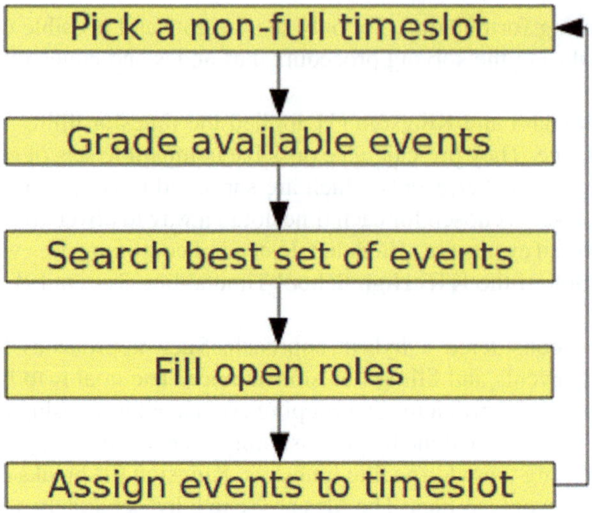

Fig. 1: Repeatedly filling timeslots

3.1.1 Constraint-Related Grades

When dealing with constraint-related grades, we try to maintain a direct connection to their weight and cost function. If assigning an event would entail a penalty by violating a soft-constraint, a negative grade of exactly this penalty will be added to the grade of this event. If an assignment helps avoiding future constraint-violations, positive grades are assigned. When having positive as well as negative grades, they are added up. The benchmark instances allow soft-constraint weights of up to 1,000. To represent the urge of avoiding hard-constraint violations—which originally do not have weights themselves—we assign such constraints a weight of 10,000. The balance between hard- and soft-constraints is maintained with a parameter *soft-constraint level*. All positive and negative grades arising from soft-constraints are multiplied with this value. This allows adjusting the influence of soft-constraints to the grades with one parameter, as we will further explain in Sect. 3.3.

We will now demonstrate the evaluation of a constraint on the example of the SpreadEventsConstraint, which is—besides the resource availability—the most important of the 15 existing constraints. First we will calculate a ratio, which expresses the urgency of an assignment independent of the weight of the constraint or whether it is hard or soft. The second step is to transform the ratio to the final grade of this constraint by considering the weights and other parameters such as the soft-constraint level.

Table 1: Variables used for calculating the SpreadEventsConstraint-ratio

Variable	Explanation
maximum	Number of allowed events per time group
cA	**C**urrent **A**ssignment: number of existing event-assignments within the current time group
pendingAssignments	nr of pending (open) event-assignments of the constrained event group
possibleAssignments	Possible assignment in all time groups (also current) without violating the *maximum*
pACG	**P**ossible **a**ssignments in **c**urrent time**g**roup permitting maximum-violations

Calculating the Ratio

The SpreadEventsConstraint is supposed to limit the usage of a set of events within certain timegroups. Typically timegroups represent days and the events form a course so that no more than one event of this course should be held on each day. The constraint allows defining a *minimum* and a *maximum* number of events per time group (day). We will only discuss the *maximum* here, where an event with a duration of more than one timeslot still counts as one assignment.

As *current time group* we understand the intersection of the time groups the constraint is applied to and the time groups the timeslot we are currently grading belongs to, which normally is the time group representing the current day (i.e., the day the timeslot we are currently grading belongs to).

Table 1 explains the variables we are going to use for calculating the ratio.

Positive grades are only applied if the maximum is not yet reached ($cA \geq maximum$). Equation (1) shows the calculation of the ratio. The left part $\frac{pendingAssignments}{possibleAssignments}$ is independent of the current time group and can be considered as a general urgency/pressure: It compares the number of pending assignments with the slots they can be assigned to without violating the maximum. The right part $\min\left(\frac{maximum-cA}{pACG},1\right)$ represents the urgency of assignments in the current time group: It relates the assignments missing to reach the maximum with the number of timeslots that are available for such assignments.

$$ratio = \frac{pendingAssignments}{possibleAssignments} \cdot \min\left(\frac{maximum - cA}{pACG}, 1\right) \tag{1}$$

If there are several possibilities ("sub-events") of assigning the duration of an event, we will calculate the ratio using the highest number of events. Having more *possibleAssignments* makes it more difficult to not violate the *maximum*, so we anticipate this case.

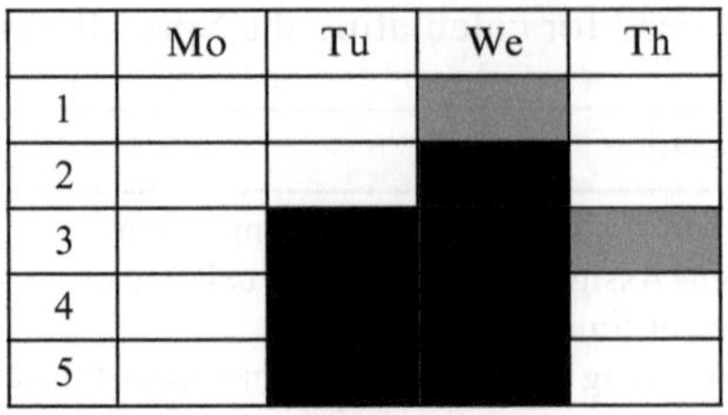

Fig. 2: Event availabilities

Example

Assume *maximum* = 1, a total number of four events, and the event availabilities as given in Fig. 2. White slots indicate that at least one of the events is available, black slots mark unavailability, and grey slots indicate that an event of this group is already assigned to the respective timeslot. The value of *pendingAssignments* is two, because two of the four events are already assigned (We-1 and Th-3). We have a *possibleAssignment* of two, one on Monday and one on Tuesday. This would imply the following ratios:

Mo-1: $ratio = \frac{2}{2} \cdot \min(\frac{1-0}{5}, 1) = 0.2$
Tu-1: $ratio = \frac{2}{2} \cdot \min(\frac{1-0}{2}, 1) = 0.5$
Th-1: $ratio = \frac{2}{2} \cdot \min(\frac{1-1}{4}, 1) = 0$

As the maximum for Th-1 is already reached with $cA \geq maximum$, we would not apply any positive grade anyhow.

Calculating the Final Grade

The final grade combines the ratio with the weight and type (soft or hard) of the constraint. It is calculated as follows:

$$\begin{aligned} \text{hard-constraint.: } grade &= ratio^{exponent} \cdot 10000 \cdot externalWeight \\ \text{soft-constraint: } grade &= ratio^{exponent} \cdot weight \cdot externalWeight \cdot softConstraintLevel \end{aligned} \tag{2}$$

Ratios usually are values between 0 and 1. The differences of a ratio can be either emphasized (stretched) by applying an *exponent* < 1, or reduced (squeezed) by *exponent* > 1. The *weight* is given by the constraint within the instance, whereas *externalWeight* is defined by us to adjust the overall importance of each constraint. These parameters will be adjusted by a hill-climbing procedure, see Sect. 3.3.

Continuing our example, assume having an external weight of 0.5, an exponent of 2, and a hard SpreadEventsConstraint. For Mo-1 this leads to $grade = 0.2^2 \cdot 10000 \cdot 0.5 = 200$, and to $grade = 0.5^2 \cdot 10000 \cdot 0.5 = 1250$ for Tu-1. When grading these timeslots, each event of the constrained event group will have the grade of the respective timeslot added.

3.1.2 Timetable-Filling Grades

To assist creating complete timetables, we calculate three more grades:

Bin-Packing When having tight resource-assignments, a problem similar to bin-packing arises. We calculate a grade that aims at gaplessly assigning such resources. This is done by favoring events and event-durations that maintain the possibility of gaplessly assigning resources.

Unassignment-Bonus Resources and events that lately failed to be assigned get an additional grade.

Course-Urgency The instances do not provide something equally to a course: Courses can be either represented by single events that have to be split or by a set of multiple events. We therefore group events that require the same resources to courses. Then, a grade is calculated favoring courses that have fewer possibilities left to assign their events, considering their duration and other constraints that impede assignments to certain timeslots.

3.2 Clique-Search

Having graded all events for the chosen timeslot, we are now looking for the most favorable set of events that can be held together. We first construct a weighted graph out of the graded events. The nodes correspond to the events having their grade as weight, and nodes are connected by edges if the events can be held simultaneously, which usually means that they do not have any resource in common. Additionally to the weight we store the *depth* of nodes, which is the number of teachers and rooms an event requires. Having open roles leads to *resource limits*, which is the number of resources of a certain type that each solution has to respect to be valid. For example, a feasible set of events must not require more gym-rooms than there are available in the given timeslot.

Apart from the resource limits and the depth, the search for the most favorable set of events corresponds to the maximum-weight clique problem. Because of the additional constraints and the instances' graphs with densities of up to 0.96% and 950 nodes, we use a custom heuristic.

A *peer* of a clique is a node that is not part of the clique but connected to all nodes of the clique. Our approach is to repeatedly *expand* cliques, which means we create new cliques by adding peers to an existing clique. When expanding a clique, we will create one new clique for each of its peers. We start at cliques of size 1, each containing one single event. Obviously, the order of selecting the clique to extend next is crucial. Condition (3) shows the basic idea of how we sort our cliques. The *depth* of a clique is the sum of the depth values of all its nodes, and the *peerDepth* of a clique is the total depth of all its peers. As long as the condition holds, a clique can still exceed the currently known maximum.

$$grade + \frac{grade}{depth} \cdot peerDepth > currentMax \tag{3}$$

This condition is incorporated into our *internGrade*, which determines the order of choosing cliques to extend. Equation (4) shows the calculation of the *internGrade*. The smaller the cliques are, the more the *internGrade* tends to over-estimate the reachable maximum, which we try to compensate with parameter c. As condition (3) aims at finding the optimum, we introduce *gradeMultiplier* (default set to 10) to focus on good-graded cliques instead of cliques having many peers. This again prevents expanding too many small cliques, and thus to expand many cliques that are highly unlikely of being further chosen later on.

$$internGrade = \left(grade \cdot gradeMultiplier + \left(peerDepth \cdot \frac{grade}{depth} \right) \right)^{1+c \cdot depth} \tag{4}$$

The cliques are stored in an array of heaps, so the *internGrade* is only calculated for the highest graded clique of each depth. At any moment, we only allow expanding cliques of certain depths, e.g. cliques that contain three to six normal-sized events. With increasing runtime, this range is shifted towards higher depths. Factor c is adapted automatically with the goal of equally choosing the cliques of the depths we currently permit; it is increased when too many cliques are chosen from low depths and decreased in case of too many cliques are chosen from high depths.

After having found a promising clique with this heuristic, the open roles have to be closed. We construct a bipartite graph out of the open roles of the chosen events and the resources that possibly fill these roles. Then, a maximum-cardinality maximum weight matching is determined. If there are resources that belong to multiple resource-groups, e.g. the sets rg_1 and rg_2, filling all open roles can be impossible when $(rg_1 \not\supseteq rg_2) \wedge (rg_1 \not\subseteq rg_2)$ and $rg_1 \cap rg_2 \neq \emptyset$. In such cases, we reduce the resource limits and repeat the search until filling all open roles succeeds. Finally, the events are assigned to the chosen timeslot.

3.3 Higher-Level Strategies

Apart from the grading parameters, we modify the search by varying how to choose the next timeslot that is filled: simply incremental or by choosing the first not entirely full timeslot of the day that has the fewest events assigned. When the pending workload of a resource gets unassignable, a local backtracking is applied. Whether or not these methods are used is controlled by parameters, too. Together with the grading parameters, they form a parameter set.

A given parameter set is tested by repeatedly applying refilling strategies, as shown in Fig. 3. We implemented several different strategies, each having a different focus: Refilling timeslots that have few events assigned, consecutively deleting and refilling timeslots or days, re-assigning resources, events or event-groups (courses) that cause high penalty or that we were unable to assign completely.

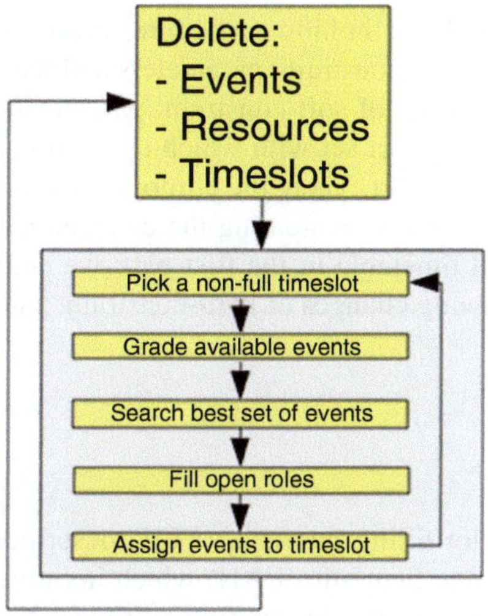

Fig. 3: Refilling strategies

After initially filling the timetable, we iteratively select and apply one of these strategies on a random basis. Hereby, each strategy has an individual selection probability which is adapted according to its success in previous applications.

On top of the refilling strategies, a hill-climbing procedure is applied. Thc goal of the hill-climbing is to find the most suitable parameter set and soft-constraint level for a given instance. To focus on the most relevant parameters at a given stage of the solving procedure, we introduced the *soft-constraint level* already mentioned in Sect. 3.1. All grades arising from soft-constraints get multiplied with this parameter. All hard-constraints and grades that assist in creating a completely filled timetable (e.g. the bin-packing and the course-urgency) are not affected by the soft-constraint level. Setting the soft-constraint level to zero completely deactivates considering any soft-constraint—the grades will exclusively aim at creating a complete timetable. Increasing the soft-constraint level leads to a higher influence of soft-constraints to the overall grades. This helps avoiding penalty, but complicates creating completely filled timetables. The hill-climbing then consists of two parts:

The first part is to find suitable parameter sets as starting points for the later hill-climbing. We try to find hard-constraint parameters that reliably fill the timetable while maximizing the extent of considering soft-constraints. To do so all soft-constraint parameters are set to default values; when creating a random parameter set these values are forbidden to be changed. Instead, all hard-constraint parameters are allowed to change. We start at a soft-constraint level of zero and create a random parameter set. This set is tested by applying the before-mentioned refilling strategies. If we achieve a valid timetable without any hard-constraint violations, we increase the soft-constraint level. Otherwise, a new random parameter set is created and evaluated.

The second part, the hill-climbing procedure, inverts the parameters that are allowed to change: The hard-constraint parameters and the soft-constraint level are fixed, we only allow changes of soft-constraint parameters at this stage. As starting point we pick the parameter set with which the lowest-penalized full timetable was created during the first part. Then, the climbing is started by iteratively changing some of the parameters and evaluating the changed parameter set. If we were not able to create a full timetable in the first part, we continue ignoring the soft-constraints by only allowing changes of hard-constraint parameters.

4 Results

We compare the timeslot-filling heuristic (TFH) described above with the more common event-assignment heuristic (EAH) which iteratively assigns the most urgent event. We simulate the latter by grading all timeslots and assigning the event that achieved the highest grade to the respective timeslot. This allows us to apply exactly the same grading procedure and higher-level strategies to both algorithms. Normally, event-assigning first picks the most urgent event (only considering hard-constraints), and then assigns it to the most suitable timeslot determined by the soft-constraints. Contrary to this we consider both the hard- and soft-constraints at the same time. We did not implement any kind of additional backtracking for EAH. The algorithms were implemented in Python. All runtimes were measured using an Intel Core 2 Duo with 2.55 GHz (with the program running on one core only) with 3 GB RAM.

We tested our algorithms on the real-world instances of the Benchmarking Project and on artificial HDTT (hard timetabling) instances from the OR Library as discussed in the next paragraphs.

4.1 Real-World Instances

Table 2 shows the best results we were able to achieve. As these are probably the first published results for the benchmarking project instances, the only source of solutions are the solutions delivered with the instances, which are listed in column *Existing Solution*. These solutions originate from previous scientific work and usually were provided by the contributor of the respective instance. The columns TFH and EAH present the results for the timeslot-filling heuristic and event-assigning heuristic, where the first value denotes the hard-constraint violations, and the latter value in parenthesis indicates the penalty arising from soft-constraint violations.

The column *runtime TFH* presents CPU-times of the initial filling of the timetable using TFH. The runtime of evaluating a parameter-set depends on the—stochastically chosen—refilling strategies. For 20 refilling-rounds, it usually is 20–30 times the initial filling time. The implementation of EAH has a lot of overhead because we

Table 2: Results for real-world instances. (best results are highlighted in boldface)

Instance	*Country*	*TFH*	*EAH*	*Existing solution*	*Runtime TFH*
Brazil1	Brazil	**0 (1)**	0 (101)	0 (104)	1.8 s
Brazil4	Brazil	**4 (1728)**	18 (1070)	–	46 s
Brazil5	Brazil	**0 (2375)**	0 (5054)	–	59 s
Brazil6	Brazil	**0 (2218)**	0 (2376)	–	40 s
Brazil7	Brazil	0 (6581)	**0 (6277)**	–	85 s
FinHigh	Finland	0 (248)	**0 (193)**	–	20 s
FinSec	Finland	**0 (216)**	0 (279)	–	40 s
FinColl	Finland	5 (424)	**4 (813)**	–	222 s
GreeceHigh	Greece	**0 (0)**	**0 (0)**	–	90 s
Patras	Greece	0 (163)	0 (30)	**0 (0)**	74 s
Preveza	Greece	0 (138)	0 (62)	**0 (0)**	68 s
Italy1	Italy	0 (138)	0 (134)	**0 (28)**	7 s
Lewitt	South-Africa	**0 (36)**	0 (144)	0 (58)	163 s
GEPRO	Netherlands	0 (19751)	36 (54157)	**1 (566)**	2,330 s
KT2003	Netherlands	0 (33565)	27 (77148)	**0 (1410)**	1,800 s
KT2005	Netherlands	23 (13530)	98 (20588)	**0 (1078)**	1,850 s
StPaul	England	0 (81996)	62 (76782)	**0 (18444)**	1,550 s

use much code that we implemented for TFH: For assigning one event, all events in all timeslots are graded without applying any delta-functions, which makes runtime-comparisons valueless.

All results were achieved using the instances of version XHSTT-2011.2. We improved two out of the nine existing solutions, but up to now we are not able to explain the rather poor results of the larger instances from the Netherlands and England. Because of the better results for smaller instances, and also for Lewitt which is comparable in size, we do not think that this is a more general problem of our approach, but rather a flaw of the grading procedure which we were unable to detect until now.

We did not include the Australian instances in our tests for two reasons: The definition of the limitWorkload constraint was problematic at the time of implementing, but is fixed already. This constraint is one of the key points of the Australian instances. Second, our algorithm is not well suited for dealing with the multitude of open roles that these instances incorporate. Closing the open roles after searching the maximum weight clique decreases control and influence on the search procedure. Having chosen a set of events, we have to fill their open roles, no matter how preferable assigning some of the resources is. The situation gets even worse when having hard avoidSplit-constraints: These define that all events of a course have to use the same resource to fill a certain open role. We already close the open role of the whole course when assigning the first event losing a lot of flexibility this way. However, this is an implementation-specific issue we became aware of too late, and not an inherent feature of our approach.

Table 3: Results for the artificial HDTT instances (best results are highlighted in boldface)

Method	*HDTT4*	*HDTT5*	*HDTT6*	*HDTT7*	*HDTT8*
SA1	–	0 (0.7)	0 (2.5)	2 (2.5)	2 (2.5)
SA2	**0 (0)**	0 (0.3)	0 (0.8)	0 (1.2)	0 (1.9)
TS	0 (0.2)	0 (2.2)	3 (5.6)	4 (10.9)	13 (17.2)
GS	5 (8.5)	11 (16.2)	19 (22.2)	26 (30.9)	29 (35.4)
NN-TT2	0 (0.1)	0 (0.5)	0 (0.8)	0 (1.1)	0 (1.4)
NN-TT3	0 (0.5)	0 (0.5)	0 (0.7)	0 (1.0)	0 (1.2)
CPMF	5 (10.7)	8 (13.2)	11 (18.7)	18 (25.6)	15 (28.6)
DWTAN	**0 (0)**	0 (0.4)	0 (1.65)	0 (2.1)	0 (3.25)
SA3	**0 (0)**	**0 (0)**	**0 (0)**	**0 (0)**	**0 (0.4)**
EAH	**0 (0)**	2 (5.4)	6 (7.9)	9 (12.0)	13 (15.1)
TFH	**0 (0)**	0 (0.6)	0 (2.1)	0 (2.5)	0 (3.1)
TFH: time	107 s	168 s	220 s	432 s	697 s

4.2 Artificial HDTT-Instances

The HDTT-Instances stem from the OR Library,[4] and are very basic: Every event has one teacher, one school-class and one room assigned. There are 30 timeslots, and every resource has to be occupied in every timeslot to get all events assigned. There are only two hard-constraints, AssignTime and AvoidClashes, and no soft-constraints. We converted these instances to the XML-format of the Benchmarking Project.

Table 3 compares our algorithms TFH and EAH with existing results: The first six methods are explained or cited in [19]. *SA* stands for Simulated Annealing, *TS* denotes Tabu Search, *GS* a Greedy Search and *NN-TT* are Hopfield Neural Networks. *DWTAN* and *CPMF* are neural network approaches too, described in [5]. The currently best method *SA3* again is a Simulated Annealing variant [11].

The first value of each cell indicates the hard-constraint violations of the best result achieved, followed by the *average* hard-constraint violations of 20 runs in parentheses. The row *TFH: time* denotes the average runtime of our TFH algorithm. On these instances, TFH yields solutions that are competitive to those of the leading existing methods, while EAH's solutions are significantly worse.

Here, we did not apply the full hill-climbing procedure. Instead, we create and test random parameter sets, which—for TFH—was sufficient to create valid solutions. Applying hill-climbing results in better average results, but causes runtimes much higher than the ones stated in existing work.

[4]http://people.brunel.ac.uk/~mastjjb/jeb/orlib/tableinfo.html.

4.3 Discussion

In general, the success of our algorithm highly depends on a suitable grading procedure. Dealing with several NP-complete subproblems, good grading functions are not easy to design and implement. Although we tried to maintain the relation between the grades and the constraints' penalties as directly as possible, the parameters for finding the best results are much more diverse than we originally expected. Despite the introduction of the soft-constraint level, the number of parameters often is still too high to lead to a clearly directed hill-climbing.

Regarding the refilling-strategies, it turned out that iteratively deleting and refilling timeslots and days helps to get all events assigned and therefore is more suitable for hard-to-fill instances. The other strategies—refilling resources, events and event-groups—help keeping the penalty low at the cost of having more difficulties assigning all events.

The results of the maximum-weight clique search are satisfying. The event-assigning heuristic EAH completely bypasses the clique search, and can therefore be used as an indicator for flaws in the clique-search. Surprisingly, especially for larger instances—where we assume that our clique-search yields worse results due to the NP-complete nature of the problem—TFH performs much better than EAH. This may partly be caused by the lack of backtracking, but it still indicates that the current bottleneck of TFH is not the clique-search, but rather the grading procedure.

Although theoretically applicable, our approach is not equally well suited for all instances. The Italian instance requires that at least one out of a set of multiple events is assigned to each of a certain group of timeslots. As TFH picks and fills single timeslots, we can either favor assigning all or none of those events. Although we can avoid assigning too many events using the resource limits of the clique-search, we lose fine-grained control at this point.

Interpreting the final results is not easy because of their diversity, and because of few existing solutions. Although the timeslot-filling heuristic TFH often yields better results than EAH, it is not justified to declare it as the better approach in general. The whole grading procedure was developed and implemented having TFH in mind. Also, the event-assigning heuristic EAH lacks backtracking. We assume that most of the existing solutions were achieved with an event-assigning approach, so the suitability of this approach for solving high-school timetabling problems is out of question. The performance of an approach strongly depends on the specifics of the given instance, as shown by the results of the artificial HDTT-instances. When developing our algorithm we did not adapt it to these instances in some more specific way. We were surprised by the good results of TFH, which are competitive to other, tailor-made algorithms. Completely filling each timeslot turned out to be much more appropriate than EAH to cope with this kind of problems. For TFH, a good solution of one timeslot occupies all resources (when not having any soft-constraints), which is exactly the bottleneck of the HDTT instances.

5 Conclusion

In this work we described a timeslot-filling heuristic (TFH) for creating high-school timetables. This heuristic is based on iteratively filling selected timeslots with sets of events. The more common approach, with which we compared our heuristic, is to iteratively assign single events. These approaches were evaluated using artificial instances as well as the real-world instances of the Benchmarking Project for (High) School Timetabling.

Although our algorithm is sometimes outperformed by tailor-made algorithms for particular instances, we demonstrated the general aptitude of the timeslot-filling heuristic. The suitability of an approach strongly depends on the characteristics of the instance it is applied to, which inhibits stating a clear winner. As the timeslot-filling approach is by far less explored, further investigation will be necessary to evaluate it in more detail. In particular, a comparison to other leading school timetabling algorithms that are flexible enough to handle the instances' constraints is required.

The main challenge definitely lies in the development of a suitable grading function. It has to maintain the balance between the various soft- and hard-constraints, between events of differing size, and the urge of creating a completely filled timetable. Future work may focus on more advanced concepts for the grading function. Self-adapting parameters or switching between various grading functions for one constraint are possible improvements. This could be done by applying local search algorithms to timetables created by the timeslot-filling heuristic. One could then focus on adapting the parameters or grading functions of the part (constraint-violation) the local search was able to improve. This would make the search for suitable parameter sets and inappropriate grading functions more efficient.

We want to thank the authors and contributors of the Benchmarking Project for their effort, and believe that their work will both help to organize and structure past and future scientific effort, and to revive the field of (real-world) high-school timetabling.

References

1. Abraham, D.J., Kingston, J.H.: Generalizing bipartite edge colouring to solve real instances of the timetabling problem. In: Burke, E.K., Causmaecker P.D. (eds.) Practice and Theory of Automated Timetabling IV, Springer Lecture Notes in Computer Science, vol. 2740, pp. 288–298. Springer, Berlin (2002)
2. Burke, E.K., Gendreau, M. (eds.): Proceedings of the 7th International Conference on the Practice and Theory of Automated Timetabling (PATAT 2008). Montreal, Canada (2008)
3. Burke, E.K., Petrovic, S.: Recent research directions in automated timetabling. Eur. J. Oper. Res. **140**(2), 266–280 (2002)
4. Burke, E.K., Rudova, H. (eds.): The Practice and Theory of Automated Timetabling VI, Lecture Notes in Computer Science, vol. 3867. Springer, Berlin (2007)

5. Carrasco, M.P., Pato, M.V.: A comparison of discrete and continuous neural network approaches to solve the class/teacher timetabling problem. Eur. J. Oper. Res. **153**(1), 65–79 (2004)
6. de Haan, P., Landman, R., Post, G., Ruizenaar, H.: A four-phase approach to a timetabling problem in secondary schools. In: Burke, E.K., Rudová, H. (eds.) Practice and Theory of Automated Timetabling VI, Lecture Notes in Computer Science, vol. 3867, pp. 423–425. Springer-Verlag, Berlin-Heidelberg, Germany (2006)
7. Even, S., Itai, A., Shamir, A.: On the complexity of timetabling and multicommodity flow problems. SIAM J. Comput. **5**, 691–703 (1976)
8. Gotlieb, C.C.: The construction of class-teacher time-tables. In: Popplewell, C.M. (ed.) Proceedings IFIP Congress 62, Information Processing, vol. 4, pp. 73–77. North-Holland Publishing (1963)
9. Kingston, J.H.: The kts high school timetabling system. In: Burke, E.K., Rudova, H. (eds.) Practice and Theory of Automated Timetabling VI, Lecture Notes in Computer Science, vol. 3867, pp. 181–195. Springer-Verlag, Berlin-Heidelberg, Germany (2006)
10. Kingston, J.H., Cooper, T.B.: The complexity of timetable construction problems. In: Burke, E.K., Ross, P. (eds.) Practice and Theory of Automated Timetabling, Lecture Notes in Computer Science, vol. 1153, pp. 283–295. Springer-Verlag, Berlin-Heidelberg, Germany (1996)
11. Liu, Y., Zhang, D., Leung, S.C.H.: A simulated annealing algorithm with a new neighborhood structure for the timetabling problem. In: GEC '09: Proceedings of the first ACM/SIGEVO Summit on Genetic and Evolutionary Computation, pp. 381–386. ACM, New York, NY, USA (2009)
12. McCollum, B., Burke, E., White, G. (eds.): Proceedings of the 8th International Conference on the Practice and Theory of Automated Timetabling (PATAT 2010). Belfast, Northern Ireland (2010)
13. Meyers, C., Orlin, J.B.: Very large-scale neighbourhood search techniques in timetabling problems. In: Practice and Theory of Automated Timetabling VI (Sixth International Conference, PATAT2006), Springer Lecture Notes in Computer Science, vol. 3867, pp. 24–39 (2007)
14. Pillay, N.: An overview of school timetabling research. In: McCollum, B., Burke, E., White, G. (eds.) Proceedings of the 8th International Conference on the Practice and Theory of Automated Timetabling (PATAT 2010), pp. 321–335 (2010)
15. Pimmer, M.: A timeslot-based heuristic approach to construct high-school timetables. Master's thesis, Vienna University of Technology (2010)
16. Post, G., Ahmadi, S., Daskalaki, S., Kingston, J., Kyngas, J., Nurmi, C., Ranson, D.: An xml format for benchmarks in high school timetabling. Ann. Oper. Res. **194**, 385–397 (2012)
17. Schaerf, A.: A survey of automated timetabling. Artif. Intell. Rev. **13**(2), 87–127 (1999)
18. Schmidt, G., Ströhlein, T.: Timetable construction - an annotated bibliography. Comput. J. **23**(4), 307–316 (1979)
19. Smith, K.A., Abramson, D., Duke, D.: Hopfield neural networks for timetabling: formulations, methods, and comparative results. Comput. Ind. Eng. **44**, 283–305 (2003)
20. Willemen, R.J.: School timetable construction: Algorithms and complexity. Ph.D. thesis, Technische Universiteit Eindhoven (2002)

A GRASP for Supply Chain Optimization with Financial Constraints per Production Unit

Sylverin Kemmoe, Philippe Lacomme, Nikolay Tchernev, and Alain Quilliot

Abstract This paper deals with supply chain optimization using the job-shop model where extra flow constraints are included modelizing cash flow exchanges between different supply chain partners including suppliers, retailers and manufacturers. The problem under study called job-shop with financial constraint is defined as a job-shop problem with simultaneous consideration of machine specific resource requirements and financial constraints. Tackling financial considerations permits to consider a proper coordination of production units when optimizing the supply chain planning and cash flows. The main goal is to obtain the smallest duration of a given supply chain operational planning that respects the budget limit of each production unit. We use a disjunctive graph formulation to model both the job-shop problem and cash flow constraints. We also propose a powerful and efficient GRASPxELS procedure as a solution method. The developed approach for supply chain modeling is novel in the sense that the model encompasses specific financial resource per machine i.e. per plant. The method efficiency is evaluated on a set of instances coming from the Laurence's job-shop instances.

1 Introduction

Supply Chain (SC) may be considered as an integrated process in which a group of several organizations, such as suppliers, producers, distributors and retailers, work

Sylverin Kemmoe
CRCGM EA 3849, Centre de Recherche Clermontois en Gestion et Management, Clermont Fd, France
e-mail: Sylverin.KEMMOE_TCHOMTE@u-clermont1.fr

Philippe Lacomme · Nikolay Tchernev · Alain Quilliot
LIMOS UMR 6158, Laboratoire d'Informatique de Modélisation et d'Optimisation des Systèmes, Aubiére, France
e-mail: lacomme, tchernev, quilliot@isima.fr

L. Di Gaspero et al. (eds.), *Advances in Metaheuristics*, Operations Research/ Computer Science Interfaces Series 53, DOI 10.1007/978-1-4614-6322-1_10,

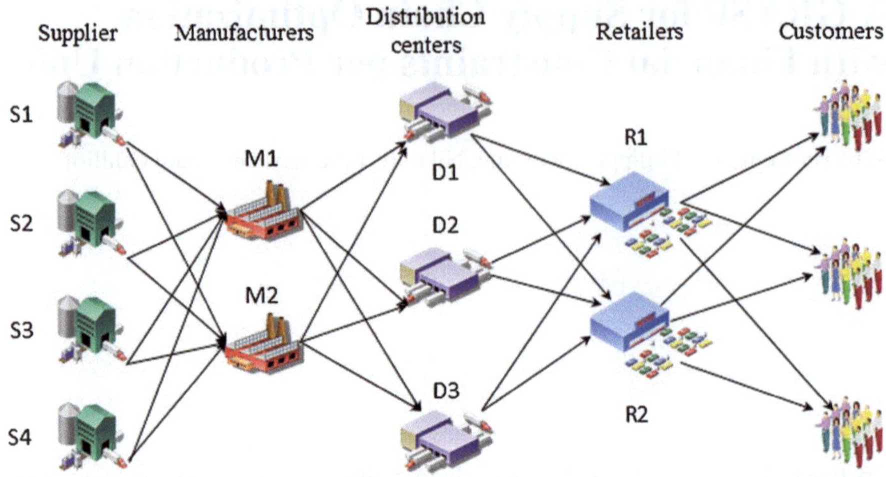

Fig. 1: Modeling supply chain

together to acquire raw materials with a view to converting them into end products which they distribute to retailers or final customers [31] (Fig. 1).

This definition stresses that all the activities along a SC should be designed according to the needs of the customers to be served. The main focus is on the order fulfillment process(es) and corresponding material, financial and information flows. Review of various supply chain classifications and models is provided by [5]. Supply chain literature often classifies supply chain flows into physical (material) flows, information flows, and cash flows [17]. Simultaneously considering supply chain production and financial problems greatly advances the efficiency of both processes. Production planning tools addresses strategic supply chain optimization including: (i) storage capacities, rules and constraints per product family required to produce; (ii) evaluating the best volume to be taken in warehouses or by manufacturers, as well as advising on minimum/maximum stocking levels at warehouses and retailers points. Scheduling problems and financial management are generally addressed sequentially: first the delivery of jobs according to due dates and warehouses capacity is addressed and then the strategic or the tactical considerations are tackled during the cash flow management. Since operational cash flows represent the result from supply chain physical flow activities, understanding the relationship between these two flows is crucial to cash flow management. To better understand, this relationship we propose to model a given supply chain (Fig. 1) as shown on Fig. 2, where each product has its own process plan which defines the product routing. Therefore the product will be treated successively by a supplier, manufacture, distribution center and retailer. Operational planning (production scheduling) has the same horizon term as cash management planning [31]. The current trend of companies is to launch customer demands as early as possible and to propose a delivery time satisfying the customer's required due date. The second-level financial decisions are constrained

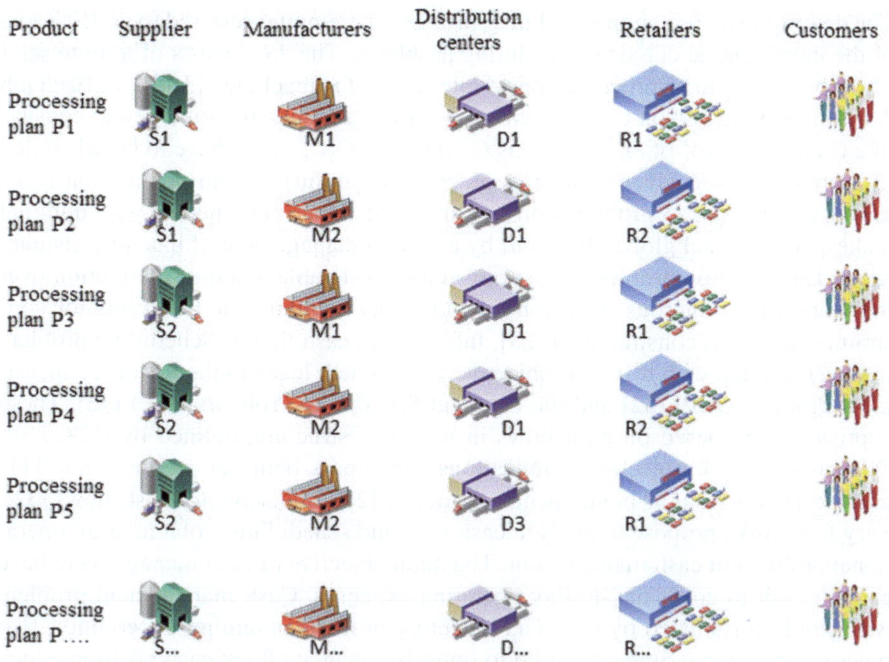

Fig. 2: Product processing plans through the supply chain

to accommodate previous scheduling solutions. It is obvious that the cash schedule and the overall spread of earnings through time, which is interrelated with the scheduling planning decisions, are optimal when both aspects are taken into account simultaneously. A series of cash flows occurs over time in two forms. Cash outflows include expenditures for labors, equipment and materials required to achieve one operation. Cash inflows are induced by progressive payment for completed work or product. This supply chain model permits to address the proper coordination between material lots (jobs) and financial considerations. This type of coordination is not a novel property but our model of the problem with financial constraints based on a disjunctive graph is a new powerful technique to obtain good solution in a few CPU times. In general scheduling decisions are made in the short term (several weeks or months) and financial decisions follow to satisfy the required base-stock levels and human resources. As shown on Fig. 2, each product has its own process plan which defines the product routing trough the supply chain.

In this work financial and cash management objectives are placed in dominant position during simultaneous financial-supply chain planning synchronization. It means that the cash position is managed by determining the best dates depending on the inflow and outflow cash dates (financial resources). There are many means to remain cash flow position positive, however our approach propose to choose the best financial exchange which minimizes the makespan taking into account material and financial constraints. Our approach is based on the job-shop theoretical model.

The deterministic job-shop scheduling problem, hereinafter referred to as JS, is one of the most generic classical scheduling problems. The JS consists of a finite set J of n jobs $\{J_i\}_{i=1}^{n}$ to be processed on a finite set M of m machines $\{M_k\}_{k=1}^{m}$: Each job J_i must be processed on every machine μ_i during a processing time p_i and consists of a chain or a set V of m_i operations $O_{i1}; O_{i2}; \ldots ; O_{im_i}$, which have to be scheduled in a predetermined given order (precedence constraint) and no preemption is allowed. Solving the JS problem consists in finding a feasible schedule with minimal makespan (minimal global duration) by a careful management of machine disjunctions. The job-shop problem has received a considerable amount of attention over the years, with numerous extensions including but not limited to transportation constraints, time-lags constraints [7, 23]. Inclusion of cash flow in scheduling problem has been studied with different objective value which leads to the Resource Investment Problem (RIP) [32] and the Payment Scheduling Problem (PSP) [33]. These approaches are based on cash flows in networks structure, defined by ([28, 29]). Depending on the objective, publications encompass both net present value [11] and extra restrictions as bonus-penalty structure [28], or discounted cash-flows [32]. Very few works propose to analyze cash flow and scheduling problem as an operational problem of cash management. The main objective of cash manager is to have enough cash to cover day-to-day operating expenses. Cash management problem was simply formulated by [4] as an inventory problem assuming uncertainty. Two types of metrics are generally used to optimize financial flow: cash position which reveals the cash which is available at the end of a specific period and cash flow which reveals cash generation during a specific period. In a recent paper, [3] optimizes financial flow and cash position in the end of each period. Forecasting a budget or the cash position consists in linking together financial and physical flows. Fenies et al. [12, 13] show the links between financial and physical flows in an tactical and operational ways, but the proposed approaches deal with a supply chain which is modelled as a flow shop. As previously stressed, there is a lack of decisional tools for supply chain optimization at the operational level taking simultaneously financial constraints and scheduling ones. Moreover, after a literature review, Elazouni and Gab-Allah [10] show that "available scheduling techniques produce financially non realistic schedules".

Some recent research papers [2, 3, 19, 20] have highlighted that financial supply chain decisions should be integrated with planning and scheduling decisions. These papers developed different mixed integer linear programming based formulations for cash management in a chemical process industry. It is evident from the above literature that the integration of financial aspects is mainly focused on scheduling and planning of batch process industries while budgeting and cash flow management are their only financial considerations. Cash management problem studied in these papers is based on maximizing the cash position by combining profit and the cost of making that profit. This approach may be applicable for manufacturing industries. Our research accepts some similarity to the approach presented in these papers. However, the paper presents a modeling and solving approach which permits to obtain positive cash position over the all scheduling horizon. Moreover, we address the problem of cash management to prioritize the operations schedule (makespan) based on incoming revenue (cash inflows) considering invoices to be paid (cash outflows). The results of our study can be applied between any two levels of upstream and downstream partners, in both manufacturing and service industry in a supply chain. The closest research to the problem we are addressing is [21], where authors have developed a mixed integer programming mode. However, the application of these models requires solving complex mathematical programming model. In this paper, we developed a modeling approach based on the JS that is much more practically relevant, and we propose a highly efficient GRASPxELS that can be applied in wide variety of situations.

This paper addresses what we call the Job Shop with Financial Constraint (JSFC) to model both expenses and final payment after jobs completion. The purpose is to find solutions which minimize the makespan by taking into account classical and financial constraints. The main goal is to obtain the smallest duration of a given supply chain operational planning that respects the budget limit of each production unit and positive cash flow. The remainder of the article is organized as follows. Section 2 introduces a GRASPxELS approach. Section 3 provides implementations for Lawrence's benchmarks, the tunning of GRASPxELS parameters and numerical experiments. The last section gives concluding remarks and further research.

2 A GRASPxELS Based Approach

2.1 GRASPxELS Description

In order to find approximate solutions to the job-shop with financial constraints, we proposed a meta-heuristic approach base on Greedy Randomized Adaptive Search Procedure (GRASP) [15, 26], hybridized with an evolutionary local search (ELS) [36], called GRASPxELS:

- GRASP is a multi-start local search metaheuristic in which each initial solution is constructed using a greedy randomized heuristic. GRASP was first proposed by Feo and Resende [14, 15]. See also Resende and Ribeiro [26] for a recent survey and [16] for a survey of a wide range of successful applications of GRASP. The multi-start approach of the GRASP provides some initial solutions, improved by a local search.
- ELS is an evolved version of iterated local search (ILS). The purpose of ELS is to better investigate the current local optimum neighborhood, before leaving it whereas the purpose of the GRASP consists in managing diversity in search space investigation. Starting from an initial solution, each ELS iteration consists in taking a copy of the incumbent solution *S*, applying a perturbation similar to the mutation operator of genetic algorithms, and improving the perturbed solution using a local search. The resulting solution *S* becomes the incumbent solution. The successive changes of encoding occur between the local search and mutations steps of our GRASPxELS.

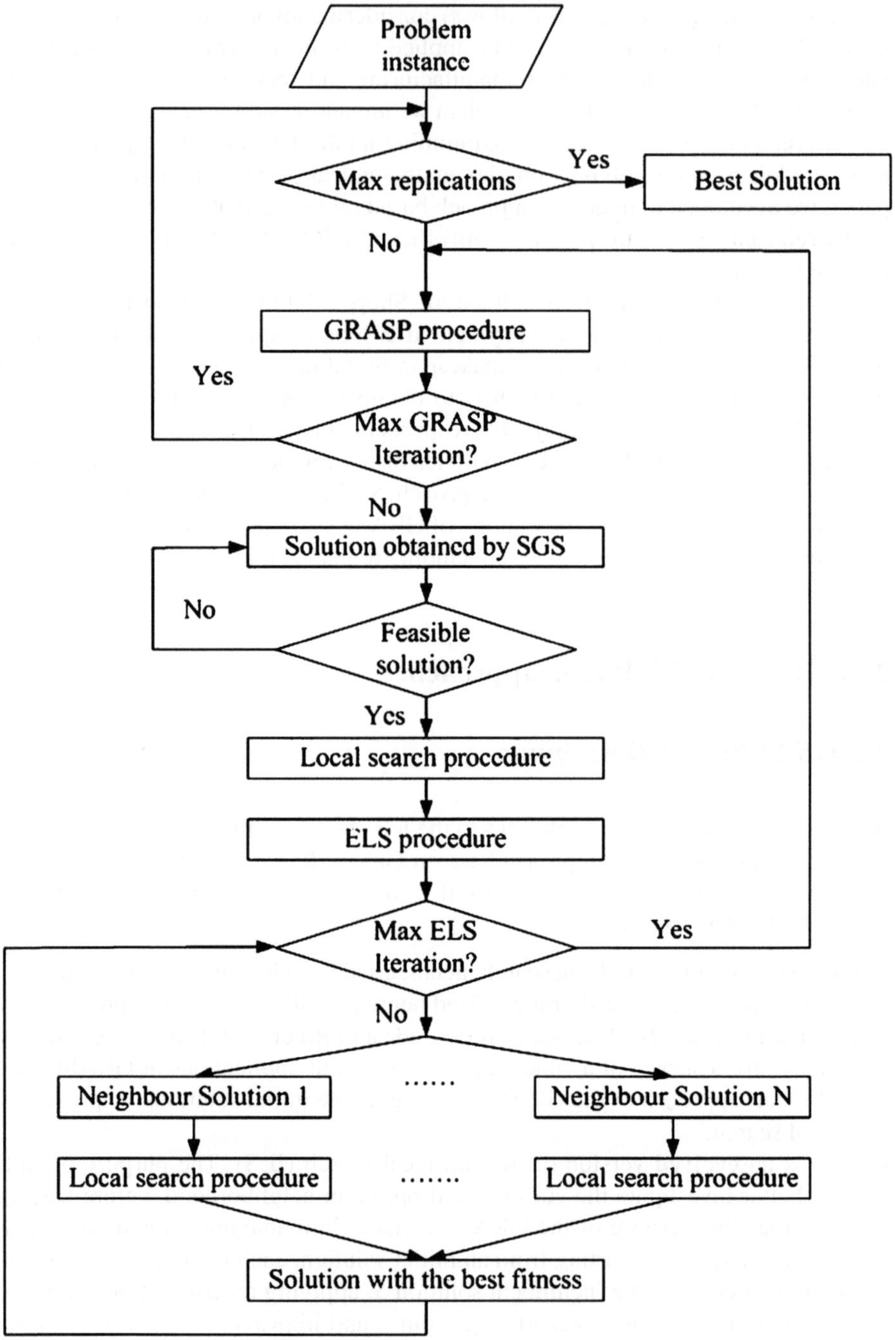

Fig. 3: GRASPxELS description

The metaheuristic approach is based on a GRASPxELS and it is illustrated in Fig. 3. The proposed GRASPxELS takes advantages of both methods. Initial solution is generated using Schedule Generation Scheme (SGS) presented is the next paragraph. Starting from SGS initial solution, the feasibility of the solution is controlled to verify that there are enough financial resources (cash flow position) to complete all operations. This solution is improved by local search to a new solution S (see Sect. 2.3.3). ELS is applied on S as follows: at each iteration N neighbour solutions are generated from the current solution S in permuting two beforehand randomly selected operations. Each neighbour solution is improved using the same local search procedure and the best solution replaces the current one S.

2.2 Scheduling Generation Scheme

Since, greedy constructive heuristics applied to solve an $NP-hard$ scheduling problem are mostly based on priority, also called dispatching rules, we choose to use a SGS based on five priority rules. A priority rule is used to select the job to be scheduled at each step of the algorithm, among the set of unscheduled jobs which are ordered in a increase or decrease orders. The used priority rules which can generate up to ten orders for machine operations are:

$Rule_1$: processing time of machine operation;
$Rule_2$: financial resource r_i^m generated by machine m for operation i (inflow)
$Rule_3$: financial resource c_i^m required to achieve on machine m operation i (outflow);
$Rule_4$: delay between the finish date of machine operation and the date of financial resource availability(inflow);
$Rule_5$: $Rule_1 + Rule_4$.

The permutation job list is obtained by one of these previous priority rules. These permutation job lists obtained by SGS are the Bierwirth's sequences of the Job Shop problem. The Bierwirth's sequence [6] introduces a job orders representation as a sequence of job numbers. Each Bierwirth's sequence denoted by λ is a job sequence in which every job i appears exactly n_i times, where n_i is the number of machine operations of job i. The first occurrence of j in the sequence refers to the first operation of the job j; the second occurrence of j in the sequence refers to the second operation of the job j; The third occurrence of j in the sequence refers to the third operation of the job j; and so on. Clearly, a Bierwirth's sequence permits to avoid inconsistent cyclic graph generation, i.e. longest path with positive cycle length can not be obtained using Bierwirth's sequence. For example if the operations i, j and k of three different jobs are processed on the same machine and if i is before j and j is before k in a Bierwirth sequence, it is mean that there is an arc from i to j and an arc from j to k. There is not an arc from j to i and from k to j. The build graph using Bierwirth's sequence is an acyclic one. Note that the number of Bierwirth's sequence is $\frac{n!}{\prod_{i=1}^{|J|} n_i!}$.

The Bierwirth's sequence has to be feasible, it means that the execution of machine operations in the order of the sequence maintain the cash flow position positive. To ensure that the Bierwirth's sequence is valid from financial point of view, the heuristic implements the following relation: $F_m = R_m^0 + \sum_{i \in V, \mu_i = m}(r_i^m - c_i^m)$, where F_m is a current cash flow position for the machine m and R_m^0 is initial cash flow position for the machine m.

To prove the feasibility of a permutation job list, the following two steps procedure is applied.

- First step: the financial amount of each machine is initialized to the initial cash flow position.
- Second step: for each operation i in Bierwirth's sequence the two following actions are performed:
  1. The selected machine is this one on which the current operation is treated.
  2. For this machine the cash position is updated $F_m = F_m - c_i^m$ and the following two conditions are verified:
     (a) If $F_m < 0$ then stop the procedure $\Longrightarrow$ unfeasible Bierwirth's sequence from financial point of view.
     (b) If $Fm \geq 0$ then F_m is updated as follows: $F_m = F_m + r_i^m$.

The heuristic that verifies the feasibility of a given sequence is given in Algorithm 1.

2.3 Bierwirth's Sequence Evaluation

The problem is modeled through simultaneous use of non oriented disjunctive graph G and on a multiflow network G^{MFN} to model the financial transfer between machine operations. A heuristic method presented above generates feasible Bierwirth's sequence λ which is used to obtain both an oriented disjunctive graph and a multiflow ϕ_λ on the G^{MFN}. Let f_{ijk} be the financial resource k exchanged between machines where operations i and j are performed. The multiflow solution is obtained by the use of multiflow heuristic procedure (Sect. 2.3.2) which gives for each couple (i, j) of machine operations the value f_{ijk}. If $f_{ijk} = 0$ then there is not a transfer of financial resource between machine operation i and machine operation j. In other words machine which performed operations i and j has enough financial resources. If $f_{ijk} \neq 0$ then the operation machine j receives from the operation machine i some financial units, it means that the starting time of the operation machine j is greater than the ending time of the operation machine i. In this case an extra arc is added. This arc has a total cost equal to the duration of the beginning operation plus the payment delay. All arcs of this type with non null flow value provided by the Bierwirth's sequence λ are added in the disjunctive graph defined by $G(\lambda, \phi_\lambda)$. The earliest starting times of machine operations in $G(\lambda, \phi_\lambda)$ can be achieved using any Bellman like longest path algorithm. Figure 4 illustrates the metaheuristic search space exploration.

Algorithm 1: Bierwirth's sequence feasibility

```
Input parameters
    λ:            Bierwirth's sequence;
    Operation:    transform Bierwirth's sequence to machine operations;
    SeqJob:       sequence of machine operations;
    Initial[k]:   initial financial resource of the machine k;
    cashflow[k]:  evolution of the financial resource of the machine k;
    cost[i, f]:   financial cost of the operation i for the supplier f;
    sell[i]:      financial counterpart of the operation i;
    μ_i:          gives the machine number of the operation i;
    K:            number of the machine;
    N:            number of the machine operations;
    F:            number of the suppliers;
Output parameters
    Response:     the boolean variable giving the feasibility of the sequence λ;
    Stop:         the boolean variable giving the stop of the procedure;
    NbOpToDo:     the number of operations remaining to be executed;
Begin
    SeqJob := Operation(λ)
    For k := 1 to K Do
        cashflow[k] := Initial[k]
    EndFor
    Stop := false;    i := 1
    While i ≤ N And Stop := false Do
        op := SeqJob[i];    k := μ_op
        For f := 1 to F Do
            cashflow[k] := cashflow[k] − cost[op, f]
        EndFor
        If cashflow[k] < 0 Then
            Stop := false
        Else
            cashflow[k] := cashflow[k] + sell[op]
        EndIf
        i := i + 1
    EndWhile
    Response := false;    NbOpToDo := N − i + 1;
    If NbOpToDo = 0 Then
        Response := true
    EndIf
End
```

This master/slave resolution search space exploration (Fig. 4) is based on a job sequences generation on machines coupled with a powerful local search procedure. The problem is first modelized through simultaneous use of a non-oriented disjunctive graph $G(\lambda)$ and a flow Graph $G^{MFN}(\lambda)$ to take into account the financial constraints. Bierwirth' sequence λ is used to obtain both an oriented disjunctive graph and a "slave" flow on the network ϕ^{λ}. The fully oriented graph $G(\lambda, \phi^{\lambda})$ encompasses both job-shop scheduling problem constraints and financial constraints.

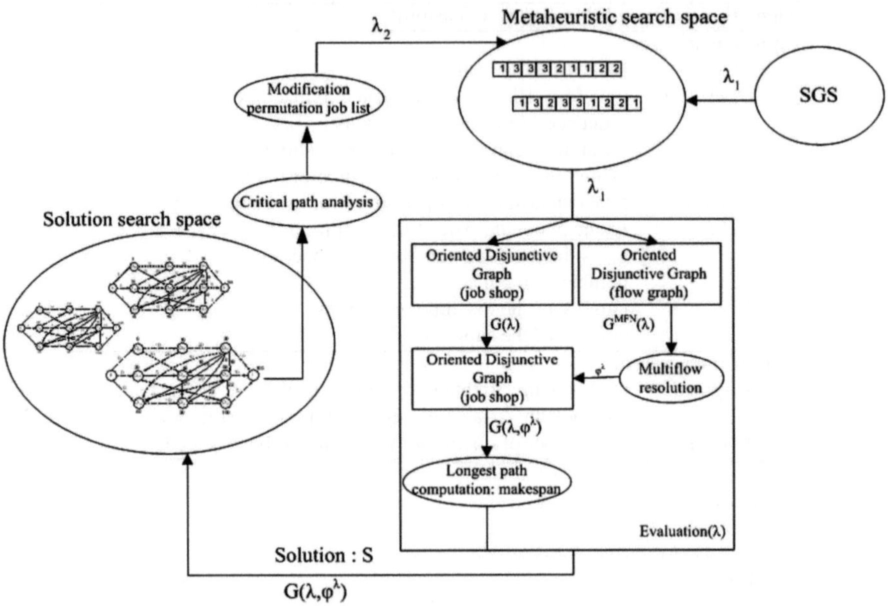

Fig. 4: Search space exploration

Table 1: A job-shop instance with financial constraints instance

J_1	$O_{11}: \quad \mu_{O_{11}}=1 \quad p_{O_{11}}=10$	$O_{12}: \quad \mu_{O_{12}}=2 \quad p_{O_{12}}=20$	$O_{13}: \quad \mu_{O_{13}}=3 \quad p_{O_{13}}=10$
	$c_{O_{11}}=20 \quad r_{O_{11}}=5$	$c_{O_{12}}=30 \quad r_{O_{12}}=15$	$c_{O_{13}}=10 \quad r_{O_{13}}=70$
J_2	$O_{21}: \quad \mu_{O_{21}}=2 \quad p_{O_{21}}=5$	$O_{22}: \quad \mu_{O_{22}}=1 \quad p_{O_{22}}=20$	$O_{23}: \quad \mu_{O_{23}}=3 \quad p_{O_{23}}=10$
	$c_{O_{21}}=25 \quad r_{O_{21}}=30$	$c_{O_{22}}=30 \quad r_{O_{22}}=45$	$c_{O_{23}}=10 \quad r_{O_{23}}=50$
J_3	$O_{31}: \quad \mu_{O_{31}}=3 \quad p_{O_{31}}=10$	$O_{32}: \quad \mu_{O_{32}}=1 \quad p_{O_{32}}=10$	$O_{33}: \quad \mu_{O_{33}}=2 \quad p_{O_{33}}=5$
	$c_{O_{31}}=40 \quad r_{O_{31}}=60$	$c_{O_{32}}=10 \quad r_{O_{32}}=70$	$c_{O_{33}}=30 \quad r_{O_{33}}=100$

Earliest starting time of operations (in $G(\lambda,\phi^{\lambda})$) can be achieved using any Bellman like longest path algorithm. At the master level, Bierwirth's sequence generation λ is devoted to a local search or ELS depending on place in the GRASPxELS algorithm.

2.3.1 Job Shop Disjunctive Graph

In this paragraph, we consider an example described in Table 1. It stresses an instance of job-shop with financial constraints. For each job the processing plan is given as well as for each machine financial data are specified in Table 1.

Figure 5 presents the corresponding non oriented disjunctive graph. The disjunctive graph, as introduced by Roy and Sussmann [27], models any job shop problem. Figure 6 is the oriented graph including the earliest starting time of operations.

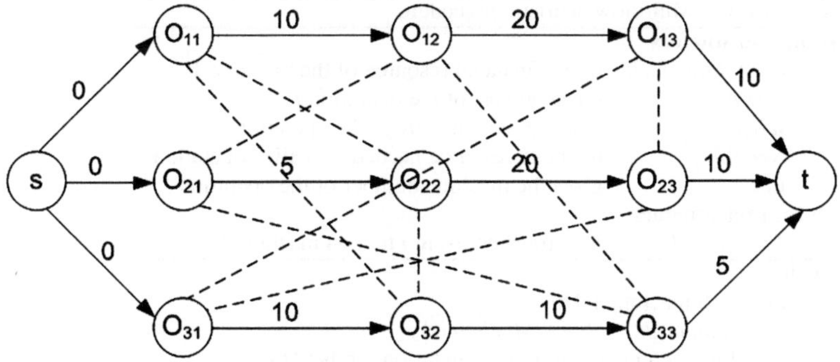

Fig. 5: Non oriented disjunctive graph G defining a job-shop problem

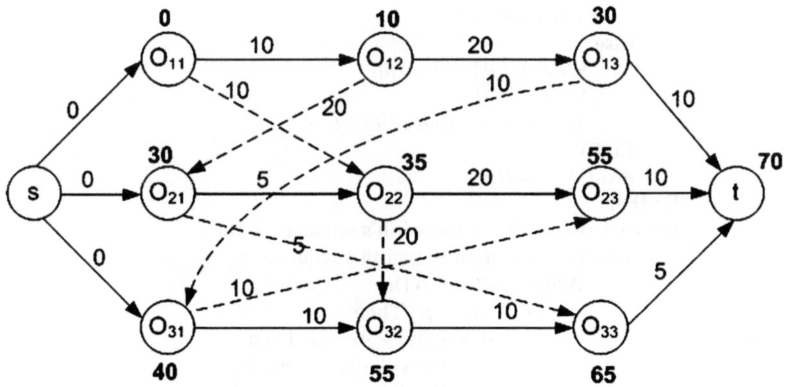

Fig. 6: Oriented disjunctive graph G defining a job-shop problem

Each job J_i has three machine operations processed on three machines μ_i during a processing time p_i. In the disjunctive graph each job operation is modeled by a vertex. Precedence constraints between operations are represented by an arc. Disjunctive constraints between two operations which require the same machine are modeled by an edge. An arc has a total cost equal to the duration of the beginning operation. Each operation i generates a financial resource (r_i) and requires financial resource (c_i). The graph with a given orientation for each edge is a conjunctive graph. Any acyclic conjunctive graph models a solution. It is obtained by defining an order of operations for each machine. Any longest path algorithm deriving from a Bellman-like algorithm permits to obtain the semi-active solution by calculating the earliest starting time of each operation.

```
Algorithm 2: Multiflow heuristic procedure
Input parameters
    InitAvMachine[k]  initial financial resource of the machine k
    cost[i]           financial cost of the operation i
    sell[i]           financial counterpart of the operation i
    Need[i]           financial resource needed to realize operation i
    μ_i               gives the machine number of the operation i
Output parameters
    Flow[i][j][k]     flow transfers from i to j on machine k
Begin
    For each machine k Do
        Amount[k] := InitAvMachine[k]
        For each operation i in permutation job list Do
            k := μ_i
            If cost[i] ≥ Amount[k] Then
                Flow[0][i][k] := Amount[k]
                Need[i] := cost[i] − Amount[k]
                Amount[k] := 0
            Else
                Flow[0][i][k] := Cost[i]
                Need[i] := 0
                Amount[k] := Amount[k] − Cost[i]
            EndIf
            Avail[i] := Sell[i]
        EndFor
        For each operation i in Bierwirth's sequence λ Do
            j the next operation Bierwirth's sequence λ
                While Avail[i] > 0 Do
                    If k = μ_i = μ_j Then
                        If Need[j] ≥ Avail[i] Then
                            Flow[i][j][k] := Avail[i]
                            Need[j] := Need[j] − Avail[i]
                            Amount[k] := 0
                        Else
                            Flow[i][j][k] := Need[j]
                            Avail[i] := Avail[i] − Need[j]
                            Need[j] := 0
                        EndIf
                    EndIf
                EndWhile
            EndFor
        EndFor
    End
```

2.3.2 Multiflow Heuristic Procedure

The principles of the multiflow heuristic are given by algorithm 2. A solution of the problem consists in computing a multiflow ϕ complying with the minimal and maximal flow constraints on the arcs. Finding a flow ϕ_λ in a graph is not straightforward and could be time consuming. Our proposal consists in computing a flow ϕ_λ in the graph $G^{MFN}(\lambda)$ which complies with the Bierwirth's sequence [6]. Such

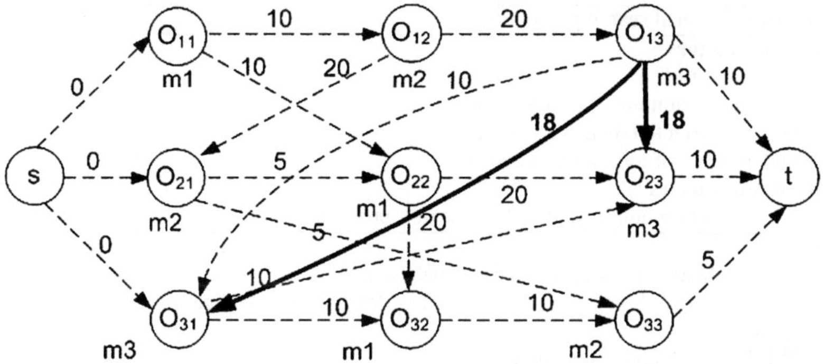

Fig. 7: Oriented disjunctive graph $G^{MFN}(\lambda, \phi_\lambda)$ defining JSFC

a flow could be denoted ϕ_λ and could be computed by any max flow algorithm used to obtain a best flow in the graph as regards the makespan. The resulting multi flow disjunctive graph with financial constraints is defined by $G^{MFN}(\lambda, \phi_\lambda)$.

The network flow has important features that are operated in a complementary manner to model the financial constraints (financial resource generated by machine operation and financial resource required by machine operation). The main idea consist in deriving flow from operation i to j according to the sequence λ. To avoid cycle, a non null flow from operation $\lambda(i)$ to $\lambda(i-k)$ is forbidden. Figure 7 gives the financial exchanges between machine operations.

For example, in Fig. 7 there is a financial transfer between two machine operations O_{13} and O_{31} which financial transfer is modeled by the arc (O_{13}, O_{31}). The delay δ of payment of each operation on machine 3 is equal to 8 units of time. The arc cost is equal to processing time of operation O_{13} plus the financial delay of payment (18=10+8). The machine operations O_{23} and O_{31} need financial resources from machine operation O_{13}. This relation is verified: $f_{O_{13},O_{21},m_3} + f_{O_{13},O_{31},m_3} \leq r_{O_{13}} = 70$.

2.3.3 Local Search

Algorithm 3 gives the principles of the local search procedure. An efficient local search is an iterative search process which tries to exploit the special structure of the problem as much as possible. The use of critical paths in disjunctive graph permits to elaborate an efficient local search for job-shop and extensions. It permits to exploit the special structure of the problem as much as possible. The proposed local search uses the concept of block approach introduced in 1986 by [18], which block approach is based on machine-block definition. The beginning and the end of two blocks are identified by two operations of different jobs on the critical path. A block approach is based on machine-block definition. The machine-block operation is a permutation between two consecutive operations on the critical path. The swap of these operations permits to generate a neighbourhood solution. The machine-block

```
Algorithm 3: Local search procedure
Input parameters
    S     a solution
    nm    maximal number of iterations
    G     non oriented disjunctive graph
    J(i)  number of job of the operation i
Output parameters
    S     a solution obtained after local search procedure
Begin
    Call Evaluation (S.Seq) to have the makespan of S
    i := P(N+1)                      //last vertex of the critical path
    g := 0
    While i ≠ 0   and   g ≤ nm Do
        j := P(i)                    //j is the father of i on the critical path
        If J_i ≠ J_j Then
            T := S;
            T.seq := S.SeqWithSwap(i, j)
            Call Evaluation (S.Seq) // to have the makespan of S
            If T.Makespan ≤ S.Makespan Then
                S := T               //new current solution and last vertex
                    i := P(N+1)      //last vertex of the critical path
            Else
                    i := P(i)        // father vertex on the critical path
            EndIf
        Else
            i := P(i)                //next vertex on the critical path
        EndIf
        g := g+1
    EndWhile
End
```

definition has been successfully applied in the very efficient taboo search [25]. [34] defined permutation between two consecutive operations on the critical path. [9] proposed an extension of the [34] neighborhood. Theoretical considerations prove that interchanging neighborhood operations on a critical path (see [34]) in the inner part of machine-block can not result in improving solutions makespan. The neighborhoods are solutions system is defined by exchanging one operation at the end of one block with another one at the beginning of the next block. The algorithm given below is based on the vertices of the critical path and on the stop criterion (maximal number of iterations noted by *nm*). The main loop consists in scanning the critical path beginning from the last operation. A neighborhood solution is obtained by the permutation of two operations of two blocks. This swap permits to obtain two situations:

- First, the new solution is better than the current one: the new solution replaces the current one $(S := T)$ and the index i is assigned to the last operation of the new critical path $(i := P(N+1))$.
- Second, the new solution is worst and the index i is updated $(i := P(i))$ to access to i father operation on the critical path.

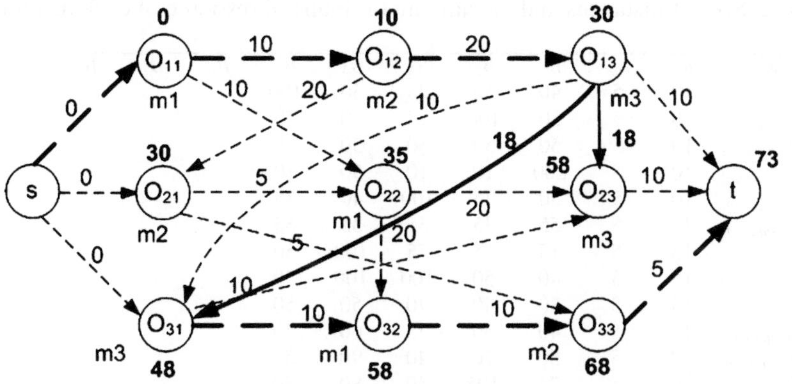

Fig. 8: An example of critical path (in bold)

Figure 8 illustrates the critical path of the job shop with financial constraints. Note that the critical path is composed either of disjunctive arcs from machines (Job-Shop constraints) or of disjunctive cash flow arcs.

3 Numerical Experiments

3.1 Implementations and Benchmarks

To test the proposed GRASPxELS we use the well known Laurence's instances (La01,...,La20) from the OR-library which instances concern classical job-shop problems. These instances have been modified to take into account the financial constraints. The instances with financial consideration (Table 2) can be downloaded at: http://www.isima.fr/lacomme/Financial_JobShop2.html. The parameters used for the instance definition are the following.

- n_J : number of jobs.
- n_m : number of machines.
- B_m : financial resource (budget) available on the machine (or production unit) m.

Table 2 gives the parameters values of each instance.

3.2 Tuning of Parameters for GRASP-ELS

3.2.1 Statistical Approach

In order to improve results of the GRASPxELS metaheuritic we use the technique proposed by [24] for parameters tuning. This statistic technique is detailed in text-

Table 2: Size of instances and the amount of financial resource of each machine B_m

Instances	n_J	n_m	B_1	B_2	B_3	B_4	B_5	B_6	B_7	B_8	B_9	B_{10}
$LA01_{Financial}$	10	5	40	50	25	30	100					
$LA02_{Financial}$	10	5	50	100	75	50	40					
$LA03_{Financial}$	10	5	50	50	50	50	50					
$LA04_{Financial}$	10	5	200	10	10	50	10					
$LA05_{Financial}$	10	5	50	75	100	40	35					
$LA06_{Financial}$	15	5	55	35	55	35	55					
$LA07_{Financial}$	15	5	15	25	75	100	50					
$LA08_{Financial}$	15	5	40	50	100	100	50					
$LA09_{Financial}$	15	5	50	40	40	50	50					
$LA10_{Financial}$	15	5	40	50	75	60	40					
$LA11_{Financial}$	20	5	50	70	40	90	50					
$LA12_{Financial}$	20	5	74	105	50	80	50					
$LA13_{Financial}$	20	5	100	50	75	50	100					
$LA14_{Financial}$	20	5	30	50	75	55	35					
$LA15_{Financial}$	20	5	50	75	100	50	59					
$LA16_{Financial}$	10	10	100	150	75	100	125	100	75	50	40	75
$LA17_{Financial}$	10	10	40	50	30	80	75	50	35	110	50	30
$LA18_{Financial}$	10	10	50	75	35	30	25	100	200	30	80	50
$LA19_{Financial}$	10	10	15	25	150	75	100	30	40	75	100	30
$LA20_{Financial}$	10	10	30	60	90	30	60	150	30	50	100	80

books like [8, 30]. For our implementation of GRASPxELS metaheuristic, we have taken 20 instances. Size of instances and the amount of financial resource of each machine B_m are given in Table 2. The GRASPxELS is executed on these instances, and for each execution, one of the three parameters (the maximum number of the GRASP iterations, the maximum number of ELS iterations, the number of neighbours generated at each ELS iteration) is modified and the GRASPxELS is run again. Each run of GRASPxELS is replicated five times.

For each run and for each replication, we note the makespan and the CPU time obtained. To compare two results, we consider that the best one is the result with the small makespan and if the makespans are equal the best result is one with the small CPU times. Each configuration is defined by the choice of a maximum number of GRASP iterations, a maximum number of ELS iterations and a number of neighbours generated.

- The maximum number of GRASP iteration : 50, 100.
- The maximum number of ELS iteration : 25, 50.
- The number of neighbour generated : 5, 10.

According to these sets, $C = 2*2*2 = 2^3 = 8$ configurations are tested in our implementation. Let us note: *GMax* the maximum number of GRASP iterations, *ELSmax* the maximum number of ELS iterations, *Neigh* the number of neighbours generated. The eight configurations $C_1, C_2, \ldots, C_8$ listed in the Table 3 have been considered for the parameters tuning.

Table 3: The eight configurations

Configuration	*GRASPmax*	*ELSmax*	*Neigh*
C_1	50	25	5
C_2	50	25	10
C_3	50	50	5
C_4	50	50	10
C_5	100	25	5
C_6	100	25	10
C_7	100	50	5
C_8	100	50	10

3.2.2 Experiments for Tuning

All procedures are implemented under Delphi 6.0 package and experiments were carried out on a 2.8 GHz computer under Windows 7 with 12 GO of memory. The results (makespan or cost and CPU time) are given for each instance and for each configuration. As we make five replications, the average value of the makespan and CPU times are considered in Table 4.

The process to determine the best configuration is done in two steps: the first test decides if the configurations are equivalent or not (hypothesis null), it does not indicate which configuration is the best. If the null hypothesis is rejected, the second test decides the best configuration. The Friedman test followed by the Bonferroni-Dunn can be used to choose among the C configurations. For each instance, the Friedman test ranks the C configuration to be executed. Let r_{ci} be the rank of the configuration c of the instance i. The mean ranks C_i are listed in Table 5 and they are graphically displayed in Fig. 9.

They seem to indicate that C_8 might be the best choice regarding solution quality, even if the metaheuristic in this configuration is the latest choice in terms of running time. Since $C = 8$ and $N = 20$ are fairly large, ($C > 4$ and $N > 15$), the Friedman test may be approximated by the chi-square distribution.

$$\chi_r^2 = \frac{12}{N*C*(C+1)} * [\sum_{c=1}^{C} (\sum_{i=1}^{N} r_{ci})^2] - 3*N*(C+1) \tag{1}$$

$$\chi_r^2(Cost) = \frac{12}{20*8*9} * [\sum_{c=1}^{8} (\sum_{i=1}^{20} r_{ci})^2] - 3*20*9$$

$$\chi_r^2(Cost) = \frac{12}{20*8*9} * [(20*8)^2 + (20*6)^2 + (20*4)^2 + (20*2.5)^2$$
$$+ (20*7)^2 + (20*5)^2 + (20*2.5)^2 + (20*1)^2] - 3*20*9 = 138.33$$

$$\chi_r^2(CPU) = \frac{12}{20*8*9} * [\sum_{c=1}^{8} (\sum_{i=1}^{20} r_{ci})^2] - 3*20*9$$

Table 4: Averages makespan (Cost) value and CPU time ($*10^{-6}$) seconds

Instance		C_1	C_2	C_3	C_4	C_5	C_6	C_7	C_8
1	*Cost*	666	666.8	666.6	666	666	666.6	666.6	666
1	*CPU*	8.1	81.9	45.1	20.9	8.7	220.1	46.5	21
2	*Cost*	664.4	660	661.2	657.6	662.2	660	659.4	657.2
2	*CPU*	97.6	162.6	224.8	582.4	209.4	172.2	455.3	667.5
3	*Cost*	670.8	667.4	666	662	667.2	665	663.6	659
3	*CPU*	106.3	149	207	318.3	147	201	290	754.5
4	*Cost*	634.4	640.6	634	636.8	632.6	637.6	619.6	626
4	*CPU*	85.1	146.2	196.8	418	152.9	251.4	484.6	911.4
5	*Cost*	598.4	598.6	598.6	598	598	598.2	598	598
5	*CPU*	31.5	32.7	15.8	111	94.6	147	109.6	141
6	*Cost*	926	926	926	926	926	926	926	926
6	*CPU*	5.1	5.4	4.9	5.5	5	5.8	4.8	6.8
7	*Cost*	891	890.2	890	890	890	890	890	890
7	*CPU*	176.6	64	50.4	81.8	297.4	226	53.4	98.4
8	*Cost*	888	888	888	888	888	888	888	888
8	*CPU*	31.5	25.4	27.1	23.2	34	29.7	26.9	26.6
9	*Cost*	951	951	951	951	951	951	951	951
9	*CPU*	101.3	49.2	51.9	26.3	108.6	52.2	55.2	31.7
10	*Cost*	958	958	958	958	958	958	958	958
10	*CPU*	3	2.5	3.1	2.6	3.3	2.6	3.2	3.2
11	*Cost*	1,222	1,222	1,222	1,222	1222	1,222	1,222	1,222
11	*CPU*	13.7	9.6	9.8	10.3	14.7	9.9	9.9	12.3
12	*Cost*	1,039	1,039	1,039	1,039	1,039	1,039	1,039	1,039
12	*CPU*	8.6	14	8.6	16.1	8.9	13.7	8.9	18.1
13	*Cost*	1,150	1,150	1,150	1,150	1,150	1,150	1,150	1,150
13	*CPU*	9.9	15.7	8.6	14.8	10.6	15.5	8.8	17.2
14	*Cost*	1,292	1,292	1,292	1,292	1,292	1,292	1,292	1,292
14	*CPU*	6.5	6.4	6.5	6.9	6.8	6.3	6.2	8.1
15	*Cost*	1,225.4	1,208.2	1,207	1,207	1,216.4	1,207	1,207	1,207
15	*CPU*	211.5	235.9	113.9	67.5	707.1	394.9	122.9	79.1
16	*Cost*	998.2	991.2	980	980	996.4	984.4	980	980
16	*CPU*	343	291	795.1	615	559.2	1,052.3	743.9	735.9
17	*Cost*	793	790	787	785.2	791.2	787.6	786.6	784.6
17	*CPU*	220.6	799.9	535.8	1,308.5	583.3	1,470.8	711.4	3,026
18	*Cost*	908	901.2	902.2	899	904.2	900	900	898
18	*CPU*	282.6	524.7	635.5	409.1	848.8	1,225.9	1,342	1,161.8
19	*Cost*	912.8	883.2	875.4	868.6	908.4	880.2	871.6	865.6
19	*CPU*	284.4	671.4	1,022.8	1,076.7	451	1,383.1	1,400.1	4,251.5
20	*Cost*	976.6	988.4	957	951.6	974.6	985.8	955.8	945.6
20	*CPU*	372	651.7	1,152.7	908	455.8	1,779.7	1,386.9	4,903.5
	Average								
	$\overline{Cost}$	918.3	915.6	912.6	911.4	916.7	914.4	911.2	910.2
	$\overline{CPU}$	119.9	197	255.8	301.1	235.4	433	363.5	843.8

$$\chi_r^2(CPU) = \frac{12}{20*8*9} * [(20*1)^2 + (20*2)^2 + (20*3)^2 + (20*4)^2$$
$$+ (20*5)^2 + (20*6)^2 + (20*7)^2 + (20*8)^2] - 3*20*9 = 140$$

Table 5: Mean rank for cost and CPU time

Instance	C_1	C_2	C_3	C_4	C_5	C_6	C_7	C_8
MeanRanks								
$\overline{Cost}$	8	6	4	2.5	7	5	2.5	1
$\overline{CPU}$	1	2	4	5	3	7	6	8
Dev to C_8								
$\overline{Cost}$	7	5	3	1.5	6	4	1.5	0
$\overline{CPU}$	−47	−6	−4	−3	−5	−1	6	0

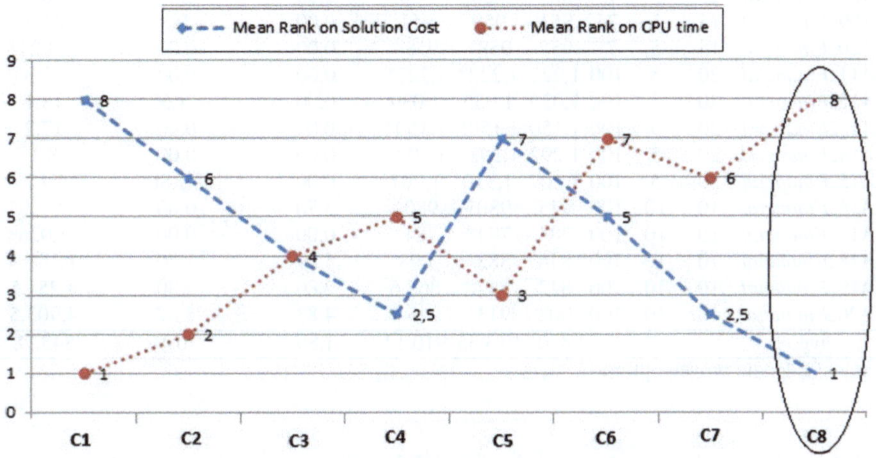

Fig. 9: Mean rank curves for the eight configurations $C_1,\ldots,C_8$

The non-parametric tests used in the sequel (sign test, Friedman test, Bonferroni-Dunn test) are detailed in textbooks like [8, 30]. The probability to reject the null hypothesis, knowing that it is true is taken at $\eta = 0.05$ for the significance level. The Friedman test rank using chi-square table returns chi-square and a p-value, i.e., the smallest significance level that would lead to rejection of the null hypothesis. The test yields $chi-square = 138.33$ and $p-value = 0.005$ for the costs, and $chi-square = 140$ and $p-value = 0.005$ for the CPU time. So we may reject the null hypothesis at the level of significance $\eta = 0.05$.

3.2.3 Results Analysis

In order to evaluate the quality of GRASPxELS results we compare them with solutions given by Integer Linear Program (ILP) model [21] solved using Cplex. The formulation for the JSFC which extends the classical mathematical model of the job-shop scheduling problem has been written by extending of both classical linear formulations of the job-shop and of the AON-flow formulation of the RCPSP

Table 6: Computational results of the problem with financial constraints

Instances	n_J	n_m	n	*LB*	*BKS*	*S*	*Dev. to LB*	*Dev. to BKS*	$T[10^{-6}\text{s}]$
LA01_Financial	10	5	50	666	666*	666*	0.00	0.00	21
LA02_Financial	10	5	50	655	655*	657.2	0.33	0.33	667.5
LA03_Financial	10	5	50	597	649*	659	10.41	1.54	754.5
LA04_Financial	10	5	50	590	646	626	6.10	−3.10	911.4
LA05_Financial	10	5	50	593	595*	598	0.84	0.50	141
LA06_Financial	15	5	75	926	926*	926*	0.00	0.00	6.8
LA07_Financial	15	5	75	890	894	890	0.00	−0.45	98.4
LA08_Financial	15	5	75	863	888	888	2.90	0.00	26.6
LA09_Financial	15	5	75	951	959	951	0.00	−0.83	31.7
LA10_Financial	15	5	75	958	958*	958*	0.00	0.00	3.2
LA11_Financial	20	5	100	1,222	1,222*	1,222*	0.00	0.00	12.3
LA12_Financial	20	5	100	1,039	1,055	1,039	0.00	−1.52	18.1
LA13_Financial	20	5	100	1,150	1,150*	1,150*	0.00	0.00	17.2
LA14_Financial	20	5	100	1,292	1,292*	1,292*	0.00	0.00	8.1
LA15_Financial	20	5	100	1,207	1,310	1,207	0.00	−7.86	79.1
LA16_Financial	10	10	100	945	980*	980*	3.70	0.00	735.9
LA17_Financial	10	10	100	784	784*	784*	0.00	0.00	3,026
LA18_Financial	10	10	100	848	883*	898	5.89	1.70	1,161.8
LA19_Financial	10	10	100	842	842*	865.6	2.80	2.80	4,251.5
LA20_Financial	10	10	100	902	913*	945.6	4.83	3.57	4,903.5
Avg.				896	913.35	910.12	1.89	−0.16	843.78

Asterisks denote proven optima

[1]. In order to take into account the financial exchange, a common practice to prepare progress payment in unit price contract is through determining the worth of the completed items based on the unit prices [22]. The results of the Linear programming and the GRASPxELS are given in Table 6. Given are n_J, number of jobs; n_m, number of machines; n, total number of operations to schedule; *LB*, the lower bound found using the best known solutions of the classical job-shop; *BKS*, best solution of the problem with financial constraints found, for a linear programming formulation given in the appendix using the CPLEX 12 solver with a time limit of 24 h (the default CPLEX configuration for all parameters including the mip branching strategy is used; *S*, the best solution found with configuration C_8 of GRASPxELS; *Dev*, the percentage deviation *S* to *LB* or to *BKS*; T, computational time (in seconds) to find *S*. In the last line *Avg* gives the averages across all instances. The experiments of the linear programs were carried out using Cplex 12 solver with a time limit upper bounded by 24 h. The ILP model is given in appendix.

The results presented in Table 6 push us into accepting that the framework has some merits with new results in job-shop instances with financial consideration. The average deviation gap is -0.16% from the best solutions found by the CPLEX solver after 24 h run time. The proposed GRASPxELS found solutions in rather short computational time (in an average of $843.78 \cdot 10^{-6}$ s, that is, in less than a second), which proves that GRASPxELS is a powerful method. Moreover, eight optimal solutions over twenty are found and four out of twelve times GRASPxELS improves over the ILP results.

4 Concluding Remarks

This article addresses the job-shop scheduling problem with financial consideration i.e. scheduling the operations efficiently while respecting the financial supply chain capacity. Case study shows the relevance of the proposed approach for a company supply chain [35], because cash flow constraint is addressed simultaneously with operational planning and scheduling. A framework based on modeling the problem as a disjunctive graph with flow consideration is introduced. A new metaheuristic which hybridizes GRASP and evolutionary local search is proposed to solve the problem. The numerical experiment proves that our framework can obtain almost optimal solutions in a rather short computational time. This work is a step forward definition of efficient models for job-shops like scheduling problems with extra cash-flow constraints. Future useful extra extension could be directed into stochastic delays in payment allowing robust solutions computation from the financial point of view [35], splitting in machine operation depending on the financial resource units, exchange and interest rates in cash flow.

Acknowledgements The authors are grateful to the anonymous referees for their helpful comments on a earlier draft of the paper.

Appendix: Mathematical Formulation [21]

Problem Parameters and Variables

The flow network model can be defined for the problem using the following notations for data:

- J : set of jobs
- M : set of machines
- S : set of suppliers
- O_j : set of machine operations of the job
- E : set of arcs such that $(i_1, i_2) \in E$ i.e. that machine operation i_2 has to start after the completion of machine operation i_1
- V : set of machine operations to schedule
- n_j : number of machine operations required by the job j
- μ_i : machine required by machine operation i
- O_{ij} : ith machine operation of the job j
- p_i : processing time of machine operation i
- r_i^m : financial resource generated by machine operation i (inflow) on the machine m
- c_{is}^m : financial resource required to achieved machine operation i (outflow) on the machine m for the supplier s

- δ_i : a delay between the achievement of machine operation i and the financial counterpart r_i
- Δ_{is} : a delay for payment for operation machine i of supplier s
- R_m^0 : is the initial cash flow for the machine m
- H : a large positive number

Linear Formulation

The linear formulation can be stated using constraints to encompass both job-shop and financial constraints.

$$
\begin{array}{lll}
\min C_{max} & & (1) \\
x_{ij} = 1 & \forall i,j \in E & (2) \\
s_j - s_i \geq p_i & \forall i,j \in E & (3) \\
s_j - s_i - H * x_{ij} \geq p_i - H + y_{ij} * \delta_i & \forall i,j \in V, \mu_i = \mu_j & (4) \\
x_{ij} + x_{ji} = 1 & \forall i,j \in V, \mu_i = \mu_j & (5) \\
\phi_{ij}^k \leq H * x_{ij} & \forall i \in V \cup \{t_1\}, \forall j \in V \cup \{t_2\}, \forall k \in M & (6) \\
\phi_{ij}^k \leq H * y_{ij} & \forall i \in V \cup \{t_1\}, \forall j \in V \cup \{t_2\}, \forall k \in M & (7) \\
y_{ij} \leq \phi_{ij}^k & \forall i \in V \cup \{t_1\}, \forall j \in V \cup \{t_2\}, \forall k \in M & (8) \\
\sum_{i \in V \cup \{t_1\}} \phi_{ij}^k = c_j^k & \forall j \in V \cup \{t_2\}, \forall k \in M & (9) \\
\sum_{j \in V \cup \{t_2\}} \phi_{ij}^k \leq r_i^k & \forall i \in V \cup \{t_1\}, \forall k \in M & (10) \\
C_{max} \geq s_i = p_i & \forall i \in V & (11) \\
x_{ij} \in \{0,1\} & \forall i \in V \cup \{t_1\}, \quad \forall j \in V \cup \{t_2\} & (12) \\
y_{ij} \in \{0,1\} & \forall i \in V \cup \{t_1\}, \quad \forall j \in V \cup \{t_2\} & (13) \\
\phi_{ij}^k \in N & \forall i \in V \cup \{t_1\}, \forall j \in V \cup \{t_2\}, \forall k \in M & (14)
\end{array}
$$

- Constraints (1) give the objective of the scheduling problem: minimization of the completion time of the last operation.
- Constraints (2) give the precedence constraint between operations. The precedence constraints encompass precedence constraints of the classical job-shop problem.
- Constraints (3) ensure that machine operation j of E starts after the completion of i i.e. $s_j \geq s_i + p_i$. Since E saved all the precedence constraint between operations of each job, constraints (3) represent conjunction between operations. When financial units are transferred from operation i to j, $x_{ij} = 1$ (thanks to constraints (6)) and the equation (3) could be rewritten $s_j - s_i \geq p_i + y_{ij} * \delta_i$ ensuring that the starting time of operation j is greater than the starting time of operation i plus the processing time of i.
- Constraints (5) are disjunctive constraints preventing that two operations processed on the same machine are not processed simultaneously.
- Constraints (6) ensure that either operation i on machine $\mu_i = \mu_j$ is processed before operation j or operation j before operation i. If there is a null financial flow from operation i to j, constraints (8) imply $y_{ij} = 0$ and constraint (6) imply $y_{ij} = 1$ if there is a non null flow from operation i to j. If $y_{ij} = 0$ and $x_{ij} = 1$

then constraint (3) can be rewritten $s_j - s_i \geq p_i$ and If $y_{ij} = 1$ and $x_{ij} = 1$ then constraint (3) can be rewritten $s_j - s_i \geq p_i + \delta_i$. If $x_{ij} = 0$, constraints (3) hold whatever y_{ij}.

- Constraints (9) state that the input flow of the operation i is equal to its financial consumption. Let us note that for a machine operation μ_i, the machine required is machine j. The machine operation consumption is represented by c_j^k where $c_j^{\mu_j} > 0$ and $c_j^k = 0$ if $k \neq \mu_j$. There only one type of flow which is non-null entering in the machine operation j.
- Constraints (10) express that the output flow of the machine operation i is less than the financial counterpart of operation i. Let us note that for a machine operation i, the machine required is machine μ_i. The machine operation counterpart is represented by r_i^k where $r_i^{\mu_i} > 0$ and $r_i^k = 0$ if $k \neq \mu_i$.
- Constraints (11) give the expression of the makespan. To comply with the classical makespan definition, the delay of payment of the operations is not included and makespan give the completion time of the last operation. Makespan definition could be extended including delay and constraints (11) could be rewritten: $\forall i \in V, C_{max} \geq s_i + p_i$.
- Constraints (12) and (13) give the domain of variables y_{ij} and x_{ij}. The proposed linear formulation encompasses both classical job-shop formulation and the RCPSP flow formulation. The linear formulation is untreatable in suitable computation time for any medium scale instances due to the numerous binary variables included in the formulation. The model presented above proves that optimal solution is induced by a simultaneous resolution of two problems: the job-shop problem and the negative/positive cash flow problem.

References

1. Artigues, C., Michelon, P., Reusser, S.: Insertion techniques for static and dynamic resource-constrained project scheduling. Eur. J. Oper. Res. **148**, 249–267 (2003)
2. Badell, M., Romero, J., Huertas, R., Puigjaner, L.: Planning, scheduling and budgeting value-added chains. Comput. Chem. Eng. **28**, 45–61 (2004)
3. Badell, M., Romero, M., Puigjaner, L.: Optimal budget and cash flows during retrofitting periods in batch chemical process industries. Int. J. Prod. Econ. **95**, 359–372 (2005)
4. Baumol, W.J.: The transaction demand for cash: an inventory approach. The Q. J. Econ. **66**, 545–556 (1952)
5. Beamon, M.: Supply chain design and analysis: models and methods. Int. J. Prod. Econ. **55**, 281–294 (1998)
6. Bierwirth, C.: A generalized permutation approach to jobshop scheduling with genetic algorithms. OR Spektrum **17**, 87–92 (1995)
7. Caumond, A., Lacomme, P., Tchernev, N.: A memetic algorithm for the job-shop with time-lags. Comput. Oper. Res. **35**(7), 2331–2356 (2008)
8. Conover, V.: Practical Nonparametric Statistics, 3rd edn. Wiley, New York (1999)
9. Dell'amico, M., Trubian, M.: Applying tabu-search to the job-shop scheduling problem. Ann. Oper. Res. **41**, 231–252 (1993)
10. Elasouni, A.M., Metwally, F.G.: Expanding finance-based scheduling to devise overalloptimized project schedules. J. Constuction Eng. M. **133**, 86–90 (2007)

11. Elmaghraby, A.E., Herroelen, W.S.: The scheduling of activities to maximize the net present value of project. Eur. J. Oper. Res. **49**, 35–49 (1990)
12. Fenies, P., Bertel, S., Roux, O.: Optimal cash flow and operational planning in a company supply chain. Int. J. Comput. Integr. Manuf. **21**(4), 440–454 (2008)
13. Fenies, P., Comelli, M., Tchernev, N.: Combined financial and physical flows evaluation for logistic process and tactical production planning: Application in a company supply chain. Int. J. Prod. Econ. **112**(1), 77–95 (2008)
14. Feo, T.A., Bard, J.: Flight scheduling and maintenance based planning. Manage. Sci. **35**(12), 1415–1432 (1989)
15. Feo, T.A., Resende, M.G.C.: Greedy randomized adaptive search procedures. J. Global Optim. **6**, 109–33 (1995)
16. Festa, P., Resende, M.G.C.: GRASP: An annotated bibliography. In: Ribeiro, C.C., Hansen, P. (eds.) Essays and Surveys on Metaheuristics, pp. 325–367. Kluwer, Dordrecht (2002)
17. Fiala, P.: Information sharing in supply chains. Omega Int. J. Manage. Sci. **33**, 419–423 (2005)
18. Grabowski, J., Nowichi, E., Zdrzalka, S.: A block approach for single machine scheduling with release dates and due dates. Eur. J. Oper. Res. **26**, 278–285 (1986)
19. Guillen, G., Badell, M., Espuna, A., Puigjaner, L.: Simultaneous optimization of process operations and financial decisions to enhance the integrated planning/scheduling of chemical supply chains. Comput. Chem. Eng. **30**(3), 421–436 (2006)
20. Guillen, G., Badell, M., Puigjaner, L.: A holistic framework for short-term supply chain management integrating production and corporate financial planning. Int. J. Prod. Econ. **106**, 288–306 (2007)
21. Kemmoe, S., Lacomme, P., Tchernev, N., Quilliot, A.: Mathematical Formulation of the supply chain with cash flow per production unit modelled as Job Shop. In: International Conference on Industrial Engineering and Systems Management, France, IESM'11, May 25–27, Metz, France (2011)
22. Lacomme, P., Kemmoe, S., Quilliot, A.: A disjunctive flow graph representation of rcpsp: a memetic algorithm based framework. Workshop International: Logistique - Transport, pages Sousse, Tunisie, 18–20 Nov. (2007)
23. Lacomme, P., Larabi, M., Tchernev, N.: Job-shop based framework for simultaneous scheduling of machines and automated guided vehicles. Int. J. Prod. Econ. Doi: 10.1016/j.ijpe.2010.07.012 online first
24. Nguyen, V., Prins, C., Prodhon, C.: Multi-Start iterated Local search with tabu search list and path relinking for the two-echelon location-routing problem. In: Eng. Appl. Artif. Intel. **25** 56–71 (2012)
25. Nowicki, E., Smutnicki, C.: A fast taboo search algorithm for the job-shop problem. Manage. Sci. **42**(26), 797–813 (1996)
26. Resende, M.G.C., Ribeiro, C.C.: GRASP with path-relinking: Recent advances and applications. In: Ibaraki, T., Nonobe, K., Yagiura, M. (eds.) Metaheuristics: Progress as Real Problem Solvers, pp. 29–63. Springer, Heidelberg (2005)
27. Roy, B., Sussmann, B.: Les problèmes d'ordonnancement avec contraintes disjonctives. Note DS no. 9 bis, SEMA, Paris (1964)
28. Russel, A.R.: A comparison of heuristics for scheduling projects with cash flows and resource restriction. Manage. Sci. **32**(10), 1291–1300 (1986)
29. Russell, A.H.: Cash flows in networks. Manage. Sci. **16**, 357–373 (1970)
30. Sheskin, D.: Handbook of Parametric and Nonparametric Statistical Procedures, 2nd edn. Chapman and Hall/CRC, Boca Raton (2000)
31. Stadtler, H.: Supply chain management and advanced planning: basics, overview and challenges. Eur. J. Oper. Res. **163**(3), 575–588 (2005)
32. Taghi, S., Najafi, A.A., Niaki, A.: A genetic algorithm for resource investment problem with discounted cash flow. Appl. Mathe. Comput. **183**, 1057–1070 (2006)
33. Ulusoy, G., Cebelli, S.: An equitable approach to the payment scheduling problem in project management. Eur. J. Oper. Res. **127**, 262–278 (2000)

34. Van Laarhoven, P.J.M., Aarts, E.H.L., Lentra, J.K.: Job-shop scheduling by simulated annealing. Oper. Res. **40** 113–125 (1992)
35. Whang, S., Lee, H.L., Padmanahbhan, V.: Information distortion in a supply chain: the bullwhip effect. Manage. Sci. **43**(4), 546–558 (1997)
36. Wolf, S., Merz, P.: Evolutionary local search for the super-peer selection problem and the p-hub median problem. In: Lecture Notes in Computer Science, vol. 4771, pp. 1–15. Springer, Berlin (2007)

MIX
Papier aus verantwortungsvollen Quellen
Paper from responsible sources
FSC® C105338

If you have any concerns about our products,
you can contact us on
ProductSafety@springernature.com

In case Publisher is established outside the EU,
the EU authorized representative is:
Springer Nature Customer Service Center GmbH
Europaplatz 3, 69115 Heidelberg, Germany

Printed by Libri Plureos GmbH
in Hamburg, Germany